Springer Theses

Recognizing Outstanding Ph.D. Research

Aims and Scope

The series "Springer Theses" brings together a selection of the very best Ph.D. theses from around the world and across the physical sciences. Nominated and endorsed by two recognized specialists, each published volume has been selected for its scientific excellence and the high impact of its contents for the pertinent field of research. For greater accessibility to non-specialists, the published versions include an extended introduction, as well as a foreword by the student's supervisor explaining the special relevance of the work for the field. As a whole, the series will provide a valuable resource both for newcomers to the research fields described, and for other scientists seeking detailed background information on special questions. Finally, it provides an accredited documentation of the valuable contributions made by today's younger generation of scientists.

Theses are accepted into the series by invited nomination only and must fulfill all of the following criteria

- They must be written in good English.
- The topic should fall within the confines of Chemistry, Physics, Earth Sciences, Engineering and related interdisciplinary fields such as Materials, Nanoscience, Chemical Engineering, Complex Systems and Biophysics.
- The work reported in the thesis must represent a significant scientific advance.
- If the thesis includes previously published material, permission to reproduce this must be gained from the respective copyright holder.
- They must have been examined and passed during the 12 months prior to nomination.
- Each thesis should include a foreword by the supervisor outlining the significance of its content.
- The theses should have a clearly defined structure including an introduction accessible to scientists not expert in that particular field.

More information about this series at http://www.springer.com/series/8790

David Hall

Discovery and Measurement of the Higgs Boson in the *WW* Decay Channel

Doctoral Thesis accepted by
the University of Oxford, UK

Author
Dr. David Hall
University of Oxford
Oxford
UK

Supervisor
Dr. Chris Hays
University of Oxford
Oxford
UK

ISSN 2190-5053 　　　　　ISSN 2190-5061　(electronic)
Springer Theses
ISBN 978-3-319-19988-7　　ISBN 978-3-319-19989-4　(eBook)
DOI 10.1007/978-3-319-19989-4

Library of Congress Control Number: 2015941337

Springer Cham Heidelberg New York Dordrecht London
© Springer International Publishing Switzerland 2015
This work is subject to copyright. All rights are reserved by the Publisher, whether the whole or part of the material is concerned, specifically the rights of translation, reprinting, reuse of illustrations, recitation, broadcasting, reproduction on microfilms or in any other physical way, and transmission or information storage and retrieval, electronic adaptation, computer software, or by similar or dissimilar methodology now known or hereafter developed.
The use of general descriptive names, registered names, trademarks, service marks, etc. in this publication does not imply, even in the absence of a specific statement, that such names are exempt from the relevant protective laws and regulations and therefore free for general use.
The publisher, the authors and the editors are safe to assume that the advice and information in this book are believed to be true and accurate at the date of publication. Neither the publisher nor the authors or the editors give a warranty, express or implied, with respect to the material contained herein or for any errors or omissions that may have been made.

Printed on acid-free paper

Springer International Publishing AG Switzerland is part of Springer Science+Business Media (www.springer.com)

By believing passionately in something that still does not exist, we create it. The nonexistent is whatever we have not sufficiently desired.

Nikos Kazantzakis

Supervisor's Foreword

The Large Hadron Collider was built with the primary goal of determining the mechanism for electroweak symmetry breaking. The leading theory, the Brout–Englert–Higgs mechanism, predicted the presence of a new scalar particle known as the Higgs boson, with couplings to other particles determined by their masses. The mass of the Higgs boson itself was unknown, but was indirectly constrained to a narrow range by other precision measurements. By the time the first beam circulated in 2008, large teams of researchers on the ATLAS and CMS experiments were preparing for the search, with data from the Tevatron collider in the United States steadily reducing the allowed Higgs boson mass range.

Less than two weeks after the first LHC beam, a faulty connection short-circuited a section of superconducting magnets, ripping them from their cement base. This incident delayed first collisions by more than a year, and to protect the LHC the centre of mass energy was cut by half. The Higgs boson production rates were thus reduced and it seemed a Higgs boson discovery would take years to achieve. However, the collider performed exceptionally well in 2011 and 2012, producing data faster than anticipated. On July 4, 2012, the ATLAS and CMS experiments announced the discovery of a new boson, with considerably less data than pre-collision predictions.

The ATLAS publication describing the discovery included three Higgs boson decay channels: $H \to \gamma\gamma$, $H \to ZZ$ and $H \to WW$. The strongest evidence for a new boson came from the $H \to \gamma\gamma$ and $H \to ZZ$ channels, for which the observed rates were somewhat higher than expected, whereas the rate in the $H \to WW$ decay channel was somewhat lower than expected. After the discovery, the LHC more than doubled the data set, and the rates in the three channels became more compatible with expectation. In early 2013, ATLAS published these rate measurements, as well as measurements demonstrating that the discovered boson was most likely a spin-0 particle—a crucial requirement for a Higgs boson.

In order to improve the precision of these measurements and that of the Higgs boson mass, ATLAS dedicated the following year to upgrading its analyses. The $H \to \gamma\gamma$ and $H \to ZZ$ channels provided the most precise measurement of the

Higgs boson mass, while the $H \to WW$ channel gave the most precise determination of the production rate. In each channel, the signal significance was above the threshold for discovery. In addition, evidence for the predicted $H \to \tau\tau$ decay was achieved.

David Hall's thesis presents the final ATLAS measurement of the $H \to WW$ decay in gluon-fusion production using the 2011 and 2012 data sets, taking the reader through the process of discovery and measurement. A precursor to the observation of the $H \to WW$ decay was the measurement of the dominant background of non-resonant WW production. This measurement shared many of the features of the $H \to WW$ measurement; one important feature was the removal or separation of events with one or more jets. A zero-jet requirement effectively eliminated the large top-quark background, but it introduced an uncertainty on the efficiency of this jet veto. David was responsible for a data-based correction to reduce this uncertainty, and for evaluating the residual uncertainty. The details of the procedure, and of the full 2011 WW cross section measurement, are provided in his thesis.

In the $H \to WW$ measurement events are separated by jet multiplicity, as events with one jet contribute significantly to the sensitivity. In the analysis, one must determine the theoretical uncertainty on the migration of events between jet bins. A covariance matrix can be constructed to describe the normalisation and migration uncertainties, and different procedures for their evaluation make different assumptions about this matrix. David studied these issues in detail and evaluated the uncertainties associated with a method that had not been previously used by the experiments, but which accommodated the highest precision calculations and thus led to a reduction in the uncertainty associated with jet binning. His thesis details this method and provides a comparison of covariance matrices for different uncertainty estimation procedures.

A final notable aspect of the thesis is the clear and detailed description of the backgrounds to the $H \to WW$ search and measurement. The dominant non-resonant WW background is normalised to data using a control region; Monte Carlo simulation is then used to model the signal region, with corresponding theoretical uncertainties. David worked directly on evaluating these uncertainties and describes them in detail. A smaller background is a W boson produced in association with an off-shell photon that splits into a pair of leptons. When only one of these leptons is reconstructed, it can mimic the leptonic decay of a W boson and serve as a background to the signal process. David performed extensive studies to ensure this process was accurately simulated, and details are given in his thesis.

Measuring the Higgs boson in the WW decay channel presents many challenges, which are apparent as one reads David's thesis. The solutions to these challenges represent an impressive body of work from many people over many years. Collecting them into a thesis with many unique insights, David provides a clear exposition of the final word on the ATLAS $H \to WW$ measurement in the discovery data set.

Oxford
March 2015

Dr. Chris Hays

Abstract

In the Standard Model of particle physics, the non-zero masses of the W and Z bosons and the fermions are generated through interactions with the Higgs field, excitations of which correspond to Higgs bosons. Thus, the experimental discovery of the Higgs boson is of prime importance to physics, and would confirm our understanding of fundamental mass generation.

This thesis describes a search for the $gg \to H \to WW \to \ell\nu\ell\nu$ process of Higgs boson production and decay. It uses the LHC Run I dataset of pp collisions recorded by the ATLAS detector, which corresponds to an integrated luminosity of 4.5 fb^{-1} at $\sqrt{s} = 7$ TeV and 20.3 fb^{-1} at $\sqrt{s} = 8$ TeV. An excess of events is observed with a significance of 4.8 standard deviations, which is consistent with Higgs boson production. The significance is extended to 6.1 standard deviations when the vector boson fusion production process is included. The measured signal strength is $1.11^{+0.23}_{-0.21}$ at $m_H = 125$ GeV. A cross section measurement of WW production, a major background to this search, is also presented using the $\sqrt{s} = 7$ TeV dataset only.

Preface

As a D.Phil. research topic, the $H \to WW$ analysis has proven to be a baptism of fire. It is the most complicated of the three "discovery channels",[1] as it involves a variety of physics objects and requires a good understanding of many difficult backgrounds. As such, the analysis took huge effort from a large number of individuals. My role focussed on theoretical aspects of the signal and background modelling, and these parts shall be emphasised. I contributed to multiple iterations of the analysis [1–8], though the version presented here is unpublished at the time of writing [9]. I also co-authored the third Yellow Report produced by the LHC Higgs Cross Section Working Group [10].

When I began the degree in October 2010, there was no direct evidence for a Higgs boson. This thesis is written from a personal perspective and motivates a low mass search by electroweak fits, when in fact this aspect was motivated later by observations of a resonance in the $\gamma\gamma$ and ZZ channels.[2] Also, an advanced search strategy is described, though the discovery of $H \to WW$ was actually a gradual process with multiple iterations of blinding, optimising and unblinding the analysis. As more data were recorded and the analysis was enhanced, the results improved.

Early on, I gained relevant insight by performing multiple WW cross section measurements [12–15]. My main contribution was a jet veto correction factor applied to the WW signal, which reduces the dominant uncertainty in the analysis. This measurement shall be described when considering the WW background to the $H \to WW$ search.

To qualify for authorship within the ATLAS collaboration, I performed Run Control shifts. I also worked within the Versatile Link project [16] to investigate radiation-hardened optical components for the HL-LHC. As this research does not easily relate to the Higgs boson, it is excluded from this thesis. However, I have published articles on the radiation tolerance of optical fibres [17] and their connectors [18].

---

[1] The $\gamma\gamma$, ZZ and WW decay channels quickly gave sensitivity to the Higgs boson ultimately discovered.

[2] Dedicated high mass searches for $H \to WW$ have also been performed [11].

Overview

This thesis describes the search, discovery and measurement of the Higgs boson using proton–proton collision data recorded by the ATLAS experiment at CERN. This is accomplished by searching for collisions where a Higgs boson is produced and subsequently decays to two W bosons, each of which decay to an electron or muon and a neutrino (i.e. $H \to WW \to \ell\nu\ell\nu$). This search suffers from large experimental backgrounds, such as continuum WW production, which must be accurately modelled to yield sensitivity to the Higgs boson.

First, the theoretical motivation for the Higgs boson is presented in Chap. 1. Then, Chap. 2 outlines some important concepts related to making precise predictions within the Standard Model, which shall be referred to throughout the thesis. The experimental setup of the LHC and the ATLAS detector are described in Chap. 3.

Focus then moves to the data analysis itself. Chapter 4 offers an overview of the entire $H \to WW$ analysis, detailing the selection of Higgs boson signal events and the rejection of backgrounds. Following this, signal modelling is described in Chap. 5, WW background modelling is described in Chap. 6 (including a dedicated cross section measurement), and the modelling of other backgrounds is described in Chap. 7. The experimental results are presented and discussed in Chap. 8. Finally, in Chaps. 9 and 10, we draw conclusions from the results of this analysis and of others conducted simultaneously at the LHC, and consider the outlook of Higgs physics.

References

1. ATLAS Collaboration, Search for the Standard Model Higgs boson in the $H \to WW \to \ell\nu\ell\nu$ decay mode with 4.7 fb^{-1} of ATLAS data at $\sqrt{s} = 7$ TeV, ATLAS-CONF-2012-012 (2012)
2. ATLAS Collaboration, Search for the Standard Model Higgs boson in the $H \to WW^{(*)} \to \ell\nu\ell\nu$ decay mode using Multivariate Techniques with 4.7 fb^{-1} of ATLAS data at $\sqrt{s} = 7$ TeV, ATLAS-CONF-2012-060 (2012)
3. ATLAS Collaboration, Search for the Standard Model Higgs boson in the $H \to WW^{(*)} \to \ell\nu\ell\nu$ decay mode with 4.7 fb^{-1} of ATLAS data at $\sqrt{s} = 7$ TeV, Phys. Lett. B **716**, 62 (2012), arXiv:1206.0756 [hep-ex]
4. ATLAS Collaboration, Observation of a new particle in the search for the Standard Model Higgs boson with the ATLAS detector at the LHC, Phys. Lett. B **716**, 1 (2012), arXiv:1207.7214 [hep-ex]
5. ATLAS Collaboration, Observation of an Excess of Events in the Search for the Standard Model Higgs Boson in the $H \to WW^{(*)} \to \ell\nu\ell\nu$ Channel with the ATLAS Detector, ATLAS-CONF-2012-098 (2012)
6. ATLAS Collaboration, Update of the $H \to WW^{(*)} \to e\nu\mu\nu$ Analysis with 13 fb^{-1} of $\sqrt{s} = 8$ TeV Data Collected with the ATLAS Detector, ATLAS-CONF-2012-158 (2012)
7. ATLAS Collaboration, Measurements of the properties of the Higgs-like boson in the $WW^{(*)} \to \ell\nu\ell\nu$ decay channel with the ATLAS detector using 25 fb^{-1} of proton–proton collision data, ATLAS-CONF-2013-030 (2013)

8. ATLAS Collaboration, Measurements of Higgs boson production and couplings in diboson final states with the ATLAS detector at the LHC, Phys. Lett. B **726**, 88 (2013), arXiv:1307.1427 [hep-ex]
9. ATLAS Collaboration, Observation and measurement of Higgs boson decays to WW^* with the ATLAS detector, (2014) (in preparation for Phys. Rev. D.)
10. LHC Higgs Cross Section Working Group, Handbook of LHC Higgs Cross Sections: 3. Higgs Properties, CERN-2013-004 (2013), arXiv:1307.1347 [hep-ph]
11. ATLAS Collaboration, Search for a high-mass Higgs boson in the $H \to WW \to \ell\nu\ell\nu$ decay channel with the ATLAS detector using 21 fb^{-1} of proton–proton collision data, ATLAS-CONF-2013-067 (2013)
12. ATLAS Collaboration, Measurement of the WW cross-section in $\sqrt{s} = 7$ TeV pp collisions with ATLAS, Phys. Rev. Lett. **107**, 041802 (2011), arXiv:1104.5225 [hep-ex]
13. ATLAS Collaboration, Measurement of the W^+W^- production cross section in proton–proton collisions at $\sqrt{s} = 7$ TeV with the ATLAS detector, ATLAS-CONF-2011-110 (2011)
14. ATLAS Collaboration, Measurement of the WW cross-section in $\sqrt{s} = 7$ TeV pp collisions with the ATLAS detector and limits on anomalous gauge couplings, Phys. Lett. B **712**, 289 (2012), arXiv:1203.6232 [hep-ex]
15. ATLAS Collaboration, Measurement of W^+W^- production in pp collisions at $\sqrt{s} = 7$ TeV with the ATLAS detector and limits on anomalous WWZ and $WW\gamma$ couplings, Phys. Rev. D **87**, 112001 (2013), arXiv:1210.2979 [hep-ex]
16. F. Vasey et al., The Versatile Link common project: feasibility report, JINST 7, C01075 (2012)
17. D. Hall, B.T. Huffman, A. Weidberg, The radiation induced attenuation of optical fibres below −20 °C exposed to lifetime HL-LHC doses at a dose rate of 700 Gy(Si)/hr, JINST 7, C01047 (2012)
18. D.C. Hall, P. Hamilton, B.T. Huffman, P.K. Teng, A.R. Weidberg, The radiation tolerance of MTP and LC optical fibre connectors to 500 kGy(Si) of gamma radiation, JINST 7, P04014 (2012)

Acknowledgements

Over the last 3.5 years, I have received help and support from many individuals. First and foremost is my supervisor, Chris Hays, whom I was fortunate to work closely with. I thank him for his patience, and for our enlightening conversations on perturbative QCD, the brewing of fine beers and how to explain the hierarchy problem with plush toys.

The ATLAS $H \to WW$ analysis group is comprised of many students, postdocs and research fellows, all collaborating to achieve a common goal. I thank Jianming Qian, Biagio Di Micco, Pierre Savard, Tatsuya Masubuchi, Christian Schmitt and Corrinne Mills for successfully convening such a complex analysis. There are too many members of the analysis team for me to mention here, but I would particularly like to thank Jonathan Long, Keisuke Yoshihara, Olivier Arnaez and Magda Chelstowska. For their help and guidance in my work on theoretical uncertainties, I thank Chris Hays, Biagio Di Micco, Justin Griffiths, Bob Kehoe and Sara Diglio. I also thank the Higgs Cross Section Working Group for their feedback on the ggF jet binning studies, in particular Andrea Banfi, Gavin Salam and Frank Tackmann. I thank MCnet for their series of summer schools, the SHERPA authors for releasing a version that enabled the $W\gamma^*$ sample to be produced, and Stefano Frixione for his advice during my time at CERN.

For the 7 TeV WW cross section measurement, I would like to thank Marc-Andre Pleier and Matthias Schott in their role as analysis convenors. I also thank Shu Li and Yusheng Wu, when we were endlessly poring over cutflows.

Within the Oxford ATLAS group, Todd Huffman and Tony Weidberg provided excellent supervision during my ATLAS service work project, and an introduction to Belgian nuclear reactors and beer (and why they should not be mixed). Thanks must also go to Gemma Wooden for getting me started in the WW and $H \to WW$ analyses at a time when she was busy finishing her D.Phil. I am truly grateful to Alex Dafinca and Lucy Kogan for sharing this experience from the start, and Jacob Howard for his overwhelming computing knowledge. I also thank Chris Y, Craig, Ellie, Jim, Kate, Mireia, Rob, Sarah and Shaun for office banter. Finally, I thank Sue Geddes and Kim Proudfoot for their help with administrative matters, and

Oxford IT support for maintaining an excellent computer farm (on which I munched through 100,000 CPU hours).

I feel very lucky to have been a member of St. Catherine's College during my D.Phil. I have met such a variety of interesting characters in the Catz MCR, particularly in my first and third years; it really has been a defining aspect of my time at Oxford University. Thanks for such great memories and I look forward to many more to come.

Of course, I also thank my parents, the rest of my family and my close friends for helping me through the difficult times.

Lastly, I thank the STFC for financially supporting me for the majority of the degree, and ACEOLE for granting me a bursary to speak at the TWEPP-11 conference in Vienna. I am also indebted to St. Catherine's College for awarding me the College Science Scholarship in my third year and a Light Senior Scholarship in my fourth year. They also provided other funding at various points throughout the degree, e.g. enabling me to attend the CLASHEP-13 school in Peru.

Contents

1	**Introduction and Theoretical Motivation**		1
	1.1	The Standard Model of Particle Physics	1
	1.2	Electroweak Unification	3
		1.2.1 The Goldstone Theorem	5
		1.2.2 The Higgs Mechanism	6
		1.2.3 Glashow-Salam-Weinberg Electroweak Theory	7
	1.3	Properties of the Higgs Boson	8
	1.4	Pre-LHC Constraints on the Higgs Boson Mass	10
		1.4.1 Direct Searches	10
		1.4.2 Precision Electroweak Fits	11
		1.4.3 Theoretical Constraints	11
	References		13
2	**Computational Techniques for the LHC**		15
	2.1	Quantum Chromodynamics	15
		2.1.1 Renormalisation and the Running Coupling Constant	15
		2.1.2 Perturbative QCD	16
		2.1.3 Resummation of Large Logarithms	17
		2.1.4 Parton Distribution Functions	18
	2.2	Monte Carlo Event Generation	19
		2.2.1 The Anatomy of an Event	19
		2.2.2 Summary of Event Generators	22
		2.2.3 Multi-leg Merging	22
		2.2.4 NLO Matching	23
		2.2.5 Additional Considerations	23
		2.2.6 Parton Shower Tuning Study	24
	2.3	Jet Algorithms	25
	References		26

3 The ATLAS Experiment ... 29
- 3.1 The Large Hadron Collider ... 29
- 3.2 pp Collision Data ... 31
  - 3.2.1 Luminosity Measurement ... 31
  - 3.2.2 Run I dataset ... 32
- 3.3 The ATLAS Detector ... 32
  - 3.3.1 Tracking ... 34
  - 3.3.2 Calorimetry ... 36
  - 3.3.3 Muon Spectrometer ... 37
  - 3.3.4 Trigger and Data Acquisition ... 38
  - 3.3.5 Detector Performance ... 39
- References ... 40

4 Overview of the $H \to WW$ Analysis ... 41
- 4.1 Experimental Signature ... 41
- 4.2 Reconstruction of Physics Objects ... 42
  - 4.2.1 Tracks ... 42
  - 4.2.2 Primary and Secondary Vertices ... 43
  - 4.2.3 Electrons ... 43
  - 4.2.4 Muons ... 48
  - 4.2.5 Jets ... 49
  - 4.2.6 b-jets ... 52
  - 4.2.7 Missing Transverse Momentum ... 52
  - 4.2.8 Object Overlap Removal ... 55
- 4.3 Event Selection Criteria ... 55
  - 4.3.1 Data Quality ... 55
  - 4.3.2 Trigger ... 56
  - 4.3.3 Pre-selection of Dilepton + p_T^{inv} Signature ... 57
  - 4.3.4 $H \to WW \to \ell\nu\ell\nu$ Decay Topology ... 59
  - 4.3.5 0-jet Selection ... 60
  - 4.3.6 1-jet Selection ... 62
  - 4.3.7 $\geq$2-jet Selection ... 64
  - 4.3.8 Summary of Signal Regions ... 67
- References ... 68

5 Signal Modelling ... 71
- 5.1 Jet-Binned Cross Sections ... 71
  - 5.1.1 Perturbative Uncertainties in Jet-Binned Cross Sections ... 71
  - 5.1.2 Combined Inclusive Prescription ... 74
  - 5.1.3 Jet Veto Efficiency Prescription ... 75
  - 5.1.4 Discussion of Results ... 78

	5.2	Monte Carlo Modelling.............................	79
		5.2.1 Higgs Boson Transverse Momentum..............	79
		5.2.2 Event Selection Acceptance	80
		5.2.3 m_T Shape Modelling	83
	References..		85
6	**WW Measurement and Modelling**		87
	6.1	Cross Section Measurement in the 0-jet Bin..............	87
		6.1.1 Reconstruction of Physics Objects................	88
		6.1.2 Event Selection Criteria......................	88
		6.1.3 Analysis Strategy	89
		6.1.4 Signal Modelling	90
		6.1.5 Background Modelling	93
		6.1.6 Experimental Results........................	96
	6.2	Background Estimation for $H \to WW$ Search.............	98
		6.2.1 Theoretical Uncertainties in α_{WW}	100
		6.2.2 m_T Shape Modelling........................	103
		6.2.3 WW Background in the ≥2-jet Bin	104
	References..		104
7	**Other Backgrounds**....................................		105
	7.1	$W+$ jet and Dijet	105
		7.1.1 The Fake Factor Method	105
		7.1.2 Lepton Anti-Identification Criteria................	107
		7.1.3 Dijet Fake Factor Measurement	107
		7.1.4 $Z+$ jet Fake Factor Measurement................	108
		7.1.5 $W+$ jet Background Estimation	109
		7.1.6 Dijet Background Estimation	111
	7.2	Non-WW Diboson................................	111
		7.2.1 Same-Sign Control Region.....................	112
		7.2.2 $W\gamma$....................................	113
		7.2.3 WZ and $W\gamma^*$	115
		7.2.4 ZZ and $Z\gamma^*$	116
	7.3	Top..	118
		7.3.1 0-jet Bin Estimation	118
		7.3.2 1-jet Bin Estimation	119
		7.3.3 ≥2-jet Bin Estimation	121
	7.4	Z/γ^*...	121
		7.4.1 Z/γ^* Boson Transverse Momentum	122
		7.4.2 $Z/\gamma^* \to \tau\tau$ Estimation	124
		7.4.3 $Z/\gamma^* \to \ell\ell$ Estimation.......................	125
	7.5	Summary of Normalisation Factors	127
	References..		128

8	**Experimental Results**		129
	8.1	Systematic Uncertainties	129
		8.1.1 Experimental Uncertainties	129
		8.1.2 Theoretical Uncertainties	131
	8.2	Statistical Model	132
		8.2.1 Discriminant Observables	132
		8.2.2 Likelihood Function	132
		8.2.3 Hypothesis Testing	135
	8.3	Results	137
		8.3.1 Exclusion, Discovery and Measurement of $gg \to H \to WW$	138
		8.3.2 Combination with VBF Analysis	138
		8.3.3 Cross Section Measurements	143
	References		144
9	**Status of Higgs Physics**		145
	9.1	Properties of the Discovered Higgs Boson	145
		9.1.1 Mass Measurement	146
		9.1.2 Coupling Measurements and Limits	146
	9.1.3	Spin and Parity Measurement	149
	9.2	Theoretical Implications	149
		9.2.1 Global Electroweak Fit	150
	9.2.2	Vacuum Stability	151
		9.2.3 The Hierarchy Problem	152
	9.3	Outlook	153
	References		154
10	**Conclusions**		157
	References		158
About the Author			159

Chapter 1
Introduction and Theoretical Motivation

The Standard Model (SM) of particle physics describes the behaviour of sub-atomic particles. Since its formulation in the 1970s, it has experienced unparalleled success in modelling a wide range of phenomena that have been experimentally verified to an extraordinary degree of precision; no experimental result within the remit of the Standard Model is currently considered to significantly contradict its validity.[1] However, there are a number of physical phenomena that the Standard Model is unable to describe: gravitational attraction between massive objects, the observed asymmetry between matter and antimatter in the Universe, and astronomical evidence for dark matter and the cosmological constant.

A crucial aspect of the SM is how non-zero masses are imparted to fundamental particles. These are forbidden by underlying symmetries of the theory, though remain an experimental fact; for example, atoms could not form if the electron did not possess mass. This is achieved via interactions with a ubiquitous Higgs field, excitations of which correspond to Higgs bosons. As the only undiscovered particle of the SM, the discovery of the Higgs boson is of utmost importance to particle physics: it would complete our knowledge of the SM, and in particular confirm the mechanism of mass generation. As such, it was a primary goal of the LHC physics program, which began in 2010.

A brief introduction to the SM is given in Sect. 1.1, outlining the particle content and interactions of the theory. In Sect. 1.2, electroweak symmetry breaking is described in detail. Then, some properties of the Higgs boson are described in Sect. 1.3, and the constraints upon its mass prior to the LHC are detailed in Sect. 1.4.

1.1 The Standard Model of Particle Physics

The SM is a gauge quantum field theory describing the kinematics and interactions of sub-atomic particles [1–6]. The dynamics of such a theory are determined by the symmetries respected by its Lagrangian. The SM is invariant under local transformations of the $SU(3) \times SU(2) \times U(1)$ gauge group, resulting in the strong, weak

---

[1] Observation of neutrino oscillations required neutrino masses to be manually added to the Standard Model. It is widely believed that their relatively small masses will be explained by new physics.

and electromagnetic forces of nature. Additionally, invariance under global transformations of the Poincaré group ensures the theory is identical in all inertial frames of reference, as asserted by special relativity.

Each constituent gauge theory of the SM describes the dynamics of a force of nature, which is mediated by a number of gauge bosons and couples to a conserved current, in accordance with Noether's theorem [7]. Quantum chromodynamics (QCD) of SU(3) describes the strong interaction, is mediated by eight gluons, and couples to colour charge. Quantum electrodynamics (QED) of U(1) describes the electromagnetic interaction, is mediated by the photon, and couples to electric charge. The weak interaction is mediated by the massive $W^{\pm}$ and Z bosons and is best understood within the context of the electroweak (EW) theory, a unification of the electromagnetic and weak interactions. A theory of gravity is not included in the SM. Significantly, the gauge groups of the strong and weak interactions are non-abelian. Physically, this means that the gauge bosons are themselves charged and therefore experience self-interactions.

The elementary particles of the SM are summarised in Fig. 1.1. They are categorised into bosons (integer spin) and fermions (half-integer spin). In addition to the gauge bosons introduced above, the Higgs boson is a by-product of electroweak symmetry breaking (described in Sect. 1.2) and couples to mass. The twelve flavours of fermions are categorised according to the interactions they experience, or equivalently the charges they possess: quarks (strong, electromagnetic, weak), charged leptons (electromagnetic, weak) and neutrinos (weak). The fermions are also arranged

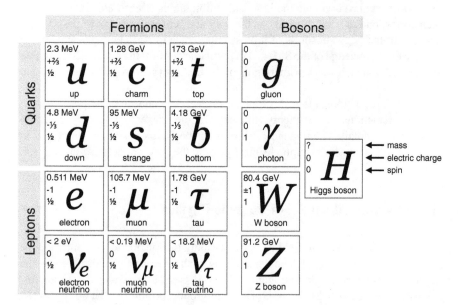

Fig. 1.1 The particle content of the SM, with masses from [8]. Constraints upon the mass of the Higgs boson are described in Sect. 1.4

1.1 The Standard Model of Particle Physics

in three generations of increasing mass. Massive particles can decay into less massive particles, while obeying the conservation laws of the SM. Fermions have an associated antiparticle with identical mass but inverted internal quantum numbers. Isolated quarks are not observed; they form colourless composite particles called hadrons.

1.2 Electroweak Unification

The first theory of weak interactions was a four-point interaction with Fermi coupling constant $G_F = 1.166 \times 10^{-5}$ GeV^{-2}. Although successful in describing low energy phenomena, such as nuclear β-decay and muon decay, at energies above $\sim$300 GeV the theory predicted cross sections which violate unitarity [1].

The solution was to introduce charged vector bosons ($W^{\pm}$ bosons) to mediate the weak interaction, similar to the exchange of photons in QED. However, unlike QED, the weak interaction is short ranged and therefore its exchange bosons must be massive. Since the propagator for a particle of mass m and momentum p contains a factor $1/(p^2 - m^2)$, in the low energy limit we can relate to Fermi's theory and identify that $G_F \sim g^2/m_W^2$, where g is the coupling of the vector boson. Thus, at low energies, the strength of the weak interaction is suppressed by the mass of the exchange boson.

At this time, there were two key obstacles to unifying the electromagnetic and weak interactions. First, the discovery of parity violation in cobalt-60 β-decay implied the weak interaction has a V–A structure, whereas QED has a pure V structure [9].[2] Second, the $W^{\pm}$ bosons are massive whilst photons are massless. This was a major problem because gauge bosons are inherently massless.[3] In fact, fermion masses were also forbidden by the chiral nature of the weak interaction, but were known to exist.[4]

Glashow's proposal of an SU(2)$_L \times$ U(1)$_Y$ group was a major step forward [3]. This model describes three gauge fields $\left(W_\mu^1, W_\mu^2, W_\mu^3\right)$ which couple to weak isospin T with strength g, and a single gauge field B_μ which couples to weak hypercharge Y with strength g'. The subscript L indicates that only left-handed chiral particles couple to the W_μ^i fields, explaining the V–A nature of the weak interaction whilst preserving QED. The physical gauge fields are obtained through the mixing of these fields

---

[2]Five bilinear covariants can be constructed from the Dirac γ matrices, which are named according to how they transform under parity: scalar, pseudoscalar, vector, axial vector and tensor.
[3]Consider the gauge transformation of a Yang-Mills gauge field $W_\mu \rightarrow W_\mu - \partial_\mu \alpha(x) - g[\alpha(x) \times W_\mu]$. Clearly the mass term $-\frac{1}{2}m^2 W_\mu \cdot W^\mu$ is not gauge invariant, and hence the gauge boson is massless.
[4]Consider a spinor as the sum of its left- and right-handed chiral states $\psi = \psi_L + \psi_R$. Then the Dirac mass term is $-m\bar{\psi}\psi = -m(\bar{\psi}_R \psi_L + \bar{\psi}_L \psi_R)$. For a chiral theory, ψ_L and ψ_R behave differently under gauge transformations and thus the mass term is not gauge invariant.

$$W^\pm_\mu = (W^1_\mu \mp i W^2_\mu)/\sqrt{2} \tag{1.1}$$
$$Z_\mu = \cos\theta_W W^3_\mu - \sin\theta_W B_\mu \tag{1.2}$$
$$A_\mu = \sin\theta_W W^3_\mu + \cos\theta_W B_\mu \tag{1.3}$$

where

$$\cos\theta_W = g/\sqrt{g^2 + g'^2} \quad\text{and}\quad \sin\theta_W = g'/\sqrt{g^2 + g'^2}. \tag{1.4}$$

We identify $W^\pm_\mu$ with the $W^\pm$ bosons, A_μ with the photon and Z_μ with a new neutral Z boson. Weak neutral currents were later confirmed experimentally [10].

Glashow's $SU(2)_L \times U(1)_Y$ theory therefore predicts the interaction of fermions, in left-handed SU(2) doublets and right-handed SU(2) singlets (see Table 1.1), with $W^\pm$, Z and γ exchange bosons. Gauge boson self-interactions are also expected due to the non-abelian nature of the EW theory. The $W^\pm$ bosons couple to weak isospin T with strength g, the Z boson couples vectorially to c_V and axially to c_A with strength $g/(2\cos\theta_W)$, and the photon couples to electric charge Q with strength $e = g\sin\theta_W$, where

$$c_V = T_3 - 2Q\sin^2\theta_W, \qquad c_A = T_3 \tag{1.5}$$
$$Q = T_3 + \frac{Y}{2}. \tag{1.6}$$

Table 1.1 The weak isospin T, weak hypercharge Y and electric charge Q of the fermions

			T	T_3	Y	Q
$\begin{pmatrix}\nu_e \\ e\end{pmatrix}_L$	$\begin{pmatrix}\nu_\mu \\ \mu\end{pmatrix}_L$	$\begin{pmatrix}\nu_\tau \\ \tau\end{pmatrix}_L$	$\frac{1}{2}$ $\frac{1}{2}$	$+\frac{1}{2}$ $-\frac{1}{2}$	-1 -1	0 -1
ν_{eR}	$\nu_{\mu R}$	$\nu_{\tau R}$	0	0	0	0
e_R	μ_R	τ_R	0	0	-2	-1
$\begin{pmatrix}u \\ d'\end{pmatrix}_L$	$\begin{pmatrix}c \\ s'\end{pmatrix}_L$	$\begin{pmatrix}t \\ b'\end{pmatrix}_L$	$\frac{1}{2}$ $\frac{1}{2}$	$+\frac{1}{2}$ $-\frac{1}{2}$	$+\frac{1}{3}$ $+\frac{1}{3}$	$+\frac{2}{3}$ $-\frac{1}{3}$
u_R	c_R	t_R	0	0	$+\frac{4}{3}$	$+\frac{2}{3}$
d_R	s_R	b_R	0	0	$-\frac{2}{3}$	$-\frac{1}{3}$

In charged currents, the states that couple to u-type quarks are superpositions of d-type quarks and are denoted with a prime. Although right-handed neutrinos are decoupled, recent observations of neutrino oscillations suggest these might exist

1.2 Electroweak Unification

Unfortunately, it was necessary to explicitly break the symmetry by adding mass terms for the $W^\pm$ and Z bosons by hand. Initial attempts to invoke a mechanism of spontaneous symmetry breaking (SSB) were hindered by the Goldstone theorem.

1.2.1 The Goldstone Theorem

SSB arises when the vacuum state does not respect the symmetry in question. This can occur when a field acquires a non-zero vacuum expectation value. To see this, consider a complex scalar field ϕ described by the Lagrangian

$$\mathcal{L} = \left(\partial_\mu \phi^\dagger\right)(\partial^\mu \phi) + \mu^2 \phi^\dagger \phi - \lambda \left(\phi^\dagger \phi\right)^2 \tag{1.7}$$

with positive μ^2 and λ, giving a sombrero potential (Fig. 1.2). Although $\mathcal{L}$ is invariant under global U(1) transformations $\phi \to e^{-i\alpha}\phi$, there are infinite degenerate vacua $\phi = \mu e^{-i\theta}/\sqrt{2\lambda}$ that are not invariant. In order to interact with the system, a single vacuum must be arbitrarily chosen, spontaneously breaking the U(1) symmetry.

The Goldstone theorem states that SSB of a continuous global symmetry will lead to the existence of a number of massless scalar Nambu-Goldstone bosons [12]. This

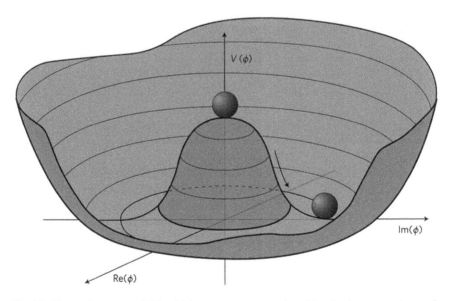

Fig. 1.2 The sombrero potential, in which a vacuum state must be arbitrarily chosen, spontaneously breaking the symmetry of the underlying Lagrangian [11]. Fluctuations in the azimuthal direction correspond to a massless Nambu-Goldstone boson. Fluctuations in the radial direction correspond to a massive Higgs boson. Reprinted by permission from Macmillan Publishers Ltd: Nature Physics **7**, 2 (2011), copyright (2011)

can be seen by considering radial and azimuthal excitations, $h(x)$ and $\theta(x)$, about the vacuum

$$\phi(x) = \frac{1}{\sqrt{2}}[v + h(x)]e^{-i\theta(x)/v} \qquad (1.8)$$

where $v = \mu/\sqrt{\lambda}$. When substituted into (1.7), we get

$$\mathcal{L} = \tfrac{1}{2}\partial_\mu\theta\partial^\mu\theta + \tfrac{1}{2}\partial_\mu h\partial^\mu h - \mu^2 h^2 + \cdots \qquad (1.9)$$

where the dots denote terms neither kinetic nor mass. We identify a massless Nambu-Goldstone boson (the θ-mode) and a Higgs boson of mass $\sqrt{2}\mu$ (the h-mode).

In order to explain massive $W^\pm$ and Z bosons, the electroweak symmetry must be broken. But the Goldstone theorem suggested that this would predict massless scalar bosons, which were not experimentally observed.

1.2.2 The Higgs Mechanism

However, when SSB of a continuous *local* symmetry is studied, something remarkable happens. The Nambu-Goldstone bosons of the theory are 'eaten' by the gauge bosons, giving them mass. The associated degrees of freedom appear as longitudinal components of the massive gauge bosons. This is known as the Higgs mechanism [13–17].

Consider the Lagrangian for a U(1) gauge theory with a sombrero potential

$$\mathcal{L} = (D_\mu\phi)^\dagger(D^\mu\phi) - \tfrac{1}{4}F_{\mu\nu}F^{\mu\nu} + \mu^2\phi^\dagger\phi - \lambda(\phi^\dagger\phi)^2 \qquad (1.10)$$

where $D_\mu = \partial_\mu + iqA_\mu$ is the covariant derivative and $F_{\mu\nu} = \partial_\mu A_\nu - \partial_\nu A_\mu$ is the field tensor. This is invariant under local U(1) transformations $\phi \to e^{-i\alpha(x)}\phi$ when accompanied by a gauge transformation of the potential $A_\mu \to A_\mu + \tfrac{1}{q}\partial_\mu\alpha(x)$.

We are free to choose the unitary gauge $\alpha(x) = -\theta(x)/v$, absorbing the θ-mode into the photon field $A_\mu \to A_\mu - \tfrac{1}{qv}\partial_\mu\theta(x)$. Ultimately, the final result is gauge-independent, but other choices require the Nambu-Goldstone bosons to be explicitly included in the Feynman rules. Since the θ-mode is 'gauged away', excitations about the vacuum become

$$\phi(x) = \frac{1}{\sqrt{2}}[v + h(x)] \qquad (1.11)$$

and the Lagrangian (1.10) becomes

$$\mathcal{L} = \tfrac{1}{2}q^2v^2 A_\mu A^\mu - \tfrac{1}{4}F_{\mu\nu}F^{\mu\nu} + \tfrac{1}{2}\partial_\mu h\partial^\mu h - \mu^2 h^2 + \cdots \qquad (1.12)$$

1.2 Electroweak Unification

where the dots denote terms neither kinetic nor mass. The Nambu-Goldstone boson is no longer present and the photon has acquired a mass qv. Again, there is a massive scalar Higgs boson as a by-product of the SSB.

1.2.3 Glashow-Salam-Weinberg Electroweak Theory

The Higgs mechanism can be extended to non-abelian gauge theories, as was necessary to describe electroweak symmetry breaking [4, 5, 18]. Consider the Lagrangian for an SU(2) × U(1) gauge theory with a sombrero potential

$$\mathcal{L} = (D_\mu \phi)^\dagger (D^\mu \phi) - \tfrac{1}{4} \boldsymbol{F}_{\mu\nu} \cdot \boldsymbol{F}^{\mu\nu} - \tfrac{1}{4} G_{\mu\nu} G^{\mu\nu} + \mu^2 \phi^\dagger \phi - \lambda (\phi^\dagger \phi)^2 \quad (1.13)$$

where $D_\mu = \partial_\mu + \tfrac{i}{2} g \boldsymbol{\tau} \cdot \boldsymbol{W}_\mu + \tfrac{i}{2} g' Y B_\mu$ is the covariant derivative, and $\boldsymbol{F}_{\mu\nu} = \partial_\mu \boldsymbol{W}_\nu - \partial_\nu \boldsymbol{W}_\mu - g \boldsymbol{W}_\mu \times \boldsymbol{W}_\nu$ and $G_{\mu\nu} = \partial_\mu B_\nu - \partial_\nu B_\mu$ are the field tensors. In this case, ϕ is an SU(2) doublet of complex scalar fields

$$\phi = \begin{pmatrix} \phi^+ \\ \phi^0 \end{pmatrix} = \frac{1}{\sqrt{2}} \begin{pmatrix} \phi_1 + i\phi_2 \\ \phi_3 + i\phi_4 \end{pmatrix}. \quad (1.14)$$

Again, there are infinite degenerate vacua satisfying $(\phi_1^2 + \phi_2^2 + \phi_3^2 + \phi_4^2) = \mu^2/\lambda$. In analogue with the abelian Higgs mechanism, the unitary gauge absorbs the ϕ_1, ϕ_2 and ϕ_4-modes into the gauge fields. Thus, considering excitations about the vacuum

$$\phi(x) = \frac{1}{\sqrt{2}} \begin{pmatrix} 0 \\ v + h(x) \end{pmatrix} \quad (1.15)$$

the Lagrangian (1.13) becomes

$$\mathcal{L} = \tfrac{1}{8} g^2 v^2 \boldsymbol{W}_\mu \cdot \boldsymbol{W}^\mu - \tfrac{1}{4} \boldsymbol{F}_{\mu\nu} \cdot \boldsymbol{F}^{\mu\nu} + \tfrac{1}{8} v^2 g'^2 B_\mu B^\mu - \tfrac{1}{4} v^2 g g' B_\mu W_3^\mu - \tfrac{1}{4} G_{\mu\nu} G^{\mu\nu}$$
$$+ \tfrac{1}{2} \partial_\mu h \partial^\mu h - \mu^2 h^2 + \cdots \quad (1.16)$$
$$= \tfrac{1}{4} g^2 v^2 W_\mu^+ W^{-\mu} - \tfrac{1}{2} (\partial_\mu W_\nu^+ - \partial_\nu W_\mu^+)(\partial^\mu W^{-\nu} - \partial^\nu W^{-\mu})$$
$$+ \tfrac{1}{8} v^2 (g^2 + g'^2) Z_\mu Z^\mu - \tfrac{1}{4} (\partial_\mu Z_\nu - \partial_\nu Z_\mu)(\partial^\mu Z^\nu - \partial^\nu Z^\mu) - \tfrac{1}{4} F_{\mu\nu} F^{\mu\nu}$$
$$+ \tfrac{1}{2} \partial_\mu h \partial^\mu h - \mu^2 h^2 + \cdots \quad (1.17)$$

where the dots denote terms neither kinetic nor mass, $F_{\mu\nu}$ is the field tensor of QED, and the expression has been rewritten in terms of the physical gauge fields using (1.1–1.3). The $W^\pm$ bosons acquire a mass $gv/2$ and the Z boson acquires a mass $v\sqrt{(g^2 + g'^2)}/2$, while the photon is massless. Again, all Nambu-Goldstone bosons are gone and a Higgs boson has appeared as a by-product of the SSB.

This theory predicts a striking relation between the gauge boson masses, using (1.4)

$$m_W = m_Z \cos \theta_W \quad (1.18)$$

which was experimentally verified once the W and Z bosons were discovered [19–23]. It also predicted a massive scalar Higgs boson, whose mass could not be determined from the other parameters of the theory.

Fermion masses can also be incorporated into EW theory through Yukawa couplings. Consider a coupling between the electron-type SU(2) doublet (see Table 1.1), the Higgs doublet ϕ given in (1.15), and the electron SU(2) singlet

$$\mathcal{L}_e^{\text{Yuk}} = -g_e(\bar{\ell}_{eL}\phi e_R + \bar{e}_R \phi^\dagger \ell_{eL}) \tag{1.19}$$

$$= -\frac{g_e}{\sqrt{2}}[v+h](\bar{e}_L e_R + \bar{e}_R e_L) \tag{1.20}$$

where g_e is the electron Yukawa coupling. The electron acquires a mass $g_e v/\sqrt{2}$ and the Higgs boson coupling to the electron is proportional to that mass (specifically m_e/v).

Finally, we note a similar phenomenon in superconductors. There, the U(1) symmetry of QED is spontaneously broken, as in Sect. 1.2.2, giving mass to the photon and thereby producing the Meissner effect. In fact, Higgs bosons have been observed in the Raman spectra of superconductors [24]. However, a major difference is that the bosonic field is a Bose-Einstein condensate of loosely bound electron pairs (known as Cooper pairs), and therefore the SSB is dynamic. This is only possible due to lattice vibrations of the underlying solid. It is natural to ask whether a similar dynamic mechanism could be used to break EW symmetry, where the Higgs boson is a composite particle. This is an active area of research, though will not be explored here.

1.3 Properties of the Higgs Boson

The Higgs boson is predicted to have zero spin and positive parity, whilst being electrically neutral and colourless. It couples directly to massive particles. Other properties, such as production cross sections and branching ratios (BRs) of decay, must be calculated as a function of its mass, which is not predicted by the SM.

At a hadron collider such as the LHC, the important production modes are gluon-gluon fusion (ggF), vector boson fusion (VBF), Higgs-strahlung (*WH* and *ZH*) and top fusion (*ttH*). Example Feynman diagrams are shown in Fig. 1.3. We note that the Higgs boson does not couple to massless gluons, therefore ggF proceeds via loops of massive coloured particles (predominantly the top quark due to its large mass).

The production cross sections at the LHC are shown in Fig. 1.4. Whilst ggF clearly dominates these rare processes, it suffers from large experimental backgrounds. The four other modes feature additional final state particles which can aid identification. For example, VBF has two well-separated quarks with no colour exchange between them.

Since the lifetime of the Higgs boson is very short, it is never directly observed in a detector. Therefore it is important to understand the BRs of its decays (Fig. 1.5). Naïvely, these are understood from the Higgs boson coupling to mass and the kine-

1.3 Properties of the Higgs Boson

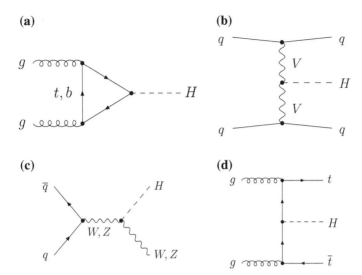

Fig. 1.3 Examples of tree-level Feynman diagrams for the Higgs production processes relevant at the LHC. **a** Gluon-gluon fusion (ggF). **b** Vector boson fusion (VBF). **c** Higgs-strahlung (*WH* and *ZH*). **d** Top fusion (*ttH*)

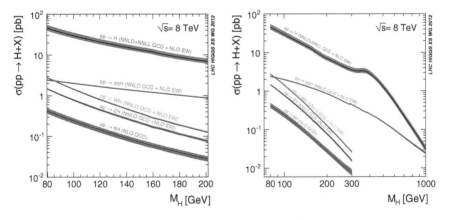

Fig. 1.4 Higgs boson production cross sections versus mass at $\sqrt{s} = 8$ TeV for a low mass range (*left*) and an expanded mass range (*right*) [25]. Theoretical uncertainties are shown as bands. The production modes are ggF (*blue*), VBF (*red*), *WH* (*green*), *ZH* (*grey*) and *ttH* (*purple*)

matic requirement $m_H > m_X + m_Y$ for a decay $H \to XY$. This is complicated by off-shell particles (*e.g* a low mass Higgs boson may decay to WW^*). Also, the $\gamma\gamma$, $Z\gamma$ and gg decay modes are different since they feature massless particles, and therefore proceed via loops of massive charged particles (electric charge for $\gamma\gamma$ and $Z\gamma$, colour charge for gg).

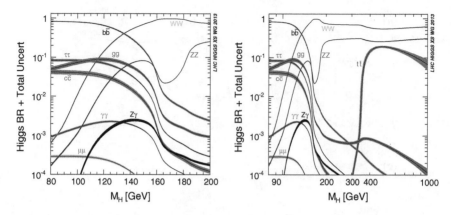

Fig. 1.5 Branching ratios of Higgs boson decay versus mass for a low mass range (*left*) and an expanded mass range (*right*) [26]. Theoretical uncertainties are shown as bands

Designing a sensitive experimental search strategy for the Higgs boson can be difficult. In decay channels featuring weak bosons, the subsequent decay of the W or Z boson must also be considered. These are more likely to decay to quarks than to leptons, but the former suffers from large backgrounds at hadron colliders. Similarly, the $b\bar{b}$ decay has the largest BR for low m_H but suffers from huge backgrounds. The sensitivity can be improved by combining with a more distinguished production mode, such as WH or ZH, but this reduces the production cross section.

1.4 Pre-LHC Constraints on the Higgs Boson Mass

Neglecting Yukawa interactions, the EW sector of the SM contains several free parameters that must be experimentally determined: two couplings (g, g') and two Higgs sector parameters (μ, λ). Using relations in Sect. 1.2, it is advantageous to choose an alternative set of independent parameters more closely connected to experiment: α_{EM}, m_W, m_Z, m_H. Finding the Higgs boson and measuring its mass is therefore of fundamental importance to understanding the EW sector, and this was a primary goal of the LHC physics program.

Prior to the LHC, the value of m_H was constrained through direct searches at previous colliders, global fits of other electroweak observables and theoretical considerations.

1.4.1 Direct Searches

Although masses below 4 GeV were excluded from B, Υ and K meson decays [27], the first meaningful searches for a Higgs boson were performed at LEP (CERN, Geneva), which ran from 1989 to 2000. This was a circular e^+e^- collider with a centre-of-mass (CM) energy tuned to the Z-pole and then later varied between 189

1.4 Pre-LHC Constraints on the Higgs Boson Mass

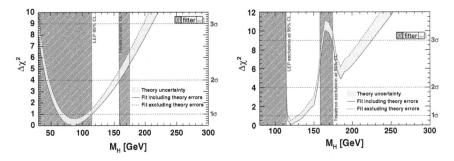

Fig. 1.6 The observed $\Delta\chi^2 = \chi^2 - \chi^2_{\min}$ of electroweak fits versus m_H, neglecting (*left*) and including (*right*) results from direct searches [30]. The exclusion limits from LEP and the Tevatron are also shown. These results were produced in July 2010. With kind permission from Springer Science and Business Media

and 209 GeV. A combined search for *ZH* was performed using a total integrated luminosity of 2.5 fb^{-1}, which excluded $m_H < 114.4$ GeV at the 95 % confidence level (CL) [28].

Further searches were performed at the Tevatron (FNAL, Illinois), which ran from 1987 to 2011. This was a circular $p\bar{p}$ collider with CM energies of 1.8 and 1.96 TeV. In 2010, searches using a variety of production and decay modes were combined across experiments using a total integrated luminosity of up to 12.6 fb^{-1}. Masses below 109 GeV and between 158 and 175 GeV were excluded at the 95 % CL [29].

1.4.2 Precision Electroweak Fits

The SM predicts that many observables will depend upon m_H through loop corrections, and it is therefore possible to infer m_H through precision EW measurements. Since the leading m_H dependence is logarithmic, the inferred constraints are weaker than those used to predict the top mass (where the dependence is quadratic).

Performing a global fit of various electroweak measurements at the LEP, SLC and Tevatron colliders (*e.g* m_W, m_Z, m_t), in July 2010 it was possible to exclude $m_H > 158$ GeV at the 95 % CL [30]. However, the best fit value was excluded by direct searches, as shown in Fig. 1.6.

1.4.3 Theoretical Constraints

Like all coupling constants in a renormalisable theory (see Sect. 2.1.1), the Higgs quartic coupling λ 'runs' with energy scale Λ, as described by the renormalisation group equations (RGEs). The running is characterised by the β-function:

$$\beta_\lambda = \frac{\partial \lambda}{\partial \log \Lambda}.$$

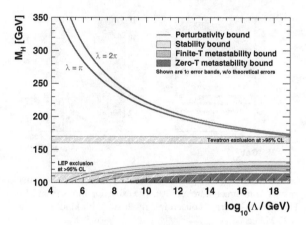

Fig. 1.7 The scale Λ at which the Higgs quartic coupling becomes non-perturbative (*blue lines*) or an instability in the EW vacuum appears (*yellow band*) [31]. The two *blue lines* represent different degrees of non-perturbativity (lower line corresponds to a two-loop correction of 25%, upper line is 50%), and their difference is indicative of the theoretical uncertainty in this bound. The *blue* and *red bands* are bounds for a metastable Universe including and neglecting thermal fluctuations respectively. Reprinted from Physics Letters B **679**, 4, J. Ellis, J.R. Espinosa, G.F. Giudice, A. Hoecker and A. Riotto, *The Probable Fate of the Standard Model*, 369–375, Copyright (2009), with permission from Elsevier

For high m_H, self-couplings dominate β_λ, which have a positive contribution. Therefore λ increases with the scale, and above some critical scale Λ_c the EW theory is no longer perturbative. Thus we would either expect to observe non-perturbative behaviour at scales $\sim \Lambda_c$ or new physics at a scale $< \Lambda_c$ that circumvents this issue. Larger values of m_H lead to lower values of Λ_c and are therefore disfavoured (blue line in Fig. 1.7). Requiring perturbativity up to the reduced Planck scale of $\bar{\Lambda}_P \sim 10^{18}$ GeV (where we expect new physics describing gravity) places an upper bound on m_H of 175 GeV [31].

For small m_H, top loops dominate β_λ, which have a negative contribution. Therefore λ decreases as the scale increases, and above some critical scale Λ_c the coupling becomes negative. Then the EW vacuum is simply a local minimum and it is possible for the Universe to collapse through quantum tunnelling into the more stable vacuum state (yellow band in Fig. 1.7). Requiring vacuum stability up to $\bar{\Lambda}_P$ places a lower bound on m_H of 129 GeV [31]. It is also possible to consider a metastable Universe whose expected lifetime is longer than its age. Accounting for thermal fluctuations up to temperatures $\sim \bar{\Lambda}_P$, the EW vacuum has a lifetime longer than the age of the Universe if $m_H > 122$ GeV (pale blue band in Fig. 1.7) [31]. These bounds are rather sensitive to the top mass, which is $m_t = 173.34 \pm 0.27(\text{stat}) \pm 0.71(\text{syst})$ GeV [32].

References

1. I.J.R. Aitchison, A.J.G. Hey, *Gauge Theories in Particle Physics*, 3rd edn. (Taylor & Francis, Abingdon, 2003)
2. M.E. Peskin, D.V. Schroeder, *An Introduction to Quantum Field Theory* (Westview Press, 1995)
3. S.L. Glashow, Partial-symmetries of weak interactions. Nucl. Phys. **22**, 579 (1961)
4. S. Weinberg, A model of Leptons. Phys. Rev. Lett. **19**, 1264 (1967)
5. A. Salam, Weak and electromagnetic interactions, in *Elementary Particle Physics:Relativistic Groups and Analyticity*, ed. by N. Svartholm. Proceedings of the Eighth Nobel Symposium (Almqvist & Wiksell, 1968), p. 367
6. G.'t Hooft, M.J.G. Veltman, Regularization and renormalization of gauge fields, Nucl. Phys. B **44**, 189 (1972)
7. E. Noether, Invariante Variationsprobleme. Nachr. v. d. Ges. d. Wiss. zu Göttingen, Math-phys. Klasse, 1918, 235 (1918)
8. Particle Data Group, Review of particle physics, Phys. Rev. D **86**, 010001 (2012), and 2013 partial update for the 2014 edn
9. C.S. Wu, E. Ambler, R.W. Hayward, D.D. Hoppes, R.P. Hudson, Experimental test of parity conservation in beta decay. Phys. Rev. **105**, 1413 (1957)
10. G. Collaboration, Observation of neutrino-like interactions without muon or electron in the gargamelle neutrino experiment. Phys. Lett. B **46**, 138 (1973)
11. L. Alvarez-Gaume, J. Ellis, Eyes on a prize particle. Nat. Phys. **7**, 2 (2011)
12. J. Goldstone, A. Salam, S. Weinberg, Broken Symmetries. Phys. Rev. **127**, 965 (1962)
13. F. Englert, R. Brout, Broken Symmetry and the Mass of Gauge Vector Mesons. Phys. Rev. Lett. **13**, 321 (1964)
14. P.W. Higgs, Broken symmetries, massless particles and gauge fields. Phys. Lett. **12**, 132 (1964)
15. P.W. Higgs, Broken symmetries and the masses of gauge bosons. Phys. Rev. Lett. **13**, 508 (1964)
16. G.S. Guralnik, C.R. Hagen, T.W.B. Kibble, Global conservation laws and massless particles. Phys. Rev. Lett. **13**, 585 (1964)
17. P.W. Higgs, Spontaneous symmetry breakdown without massless bosons. Phys. Rev. **145**, 1156 (1966)
18. T.W.B. Kibble, Symmetry breaking in non-abelian gauge theories. Phys. Rev. **155**, 1554 (1967)
19. UA1 Collaboration, Experimental observation of isolated large transverse energy electrons with associated missing energy at $\sqrt{s} = 540$ GeV. Phys. Lett. B **122**, 103 (1983)
20. UA2 Collaboration, Observation of single isolated electrons of high transverse momentum in events with missing transverse energy at the CERN $\bar{p}p$ collider. Phys. Lett. B **122**, 476 (1983)
21. UA1 Collaboration, Experimental observation of Lepton pairs of invariant mass around 95 GeV/c^2 at the CERN SPS collider. Phys. Lett. B **126**, 398 (1983)
22. UA2 Collaboration, Evidence for $Z^0 \to e^+e^-$ at the CERN $\bar{p}p$ collider. Phys. Lett. B **129**, 130 (1983)
23. UA1 Collaboration, Studies of intermediate vector boson production and decay in UA1 at the CERN proton-antiproton collider. Z. Phys. C **44**, 15 (1989)
24. P.B. Littlewood, C.M. Varma, Gauge-invariant theory of the dynamical interaction of charge density waves and superconductivity. Phys. Rev. Lett. **47**, 811 (1981)
25. LHC Higgs Cross Section Working Group, Handbook of LHC Higgs Cross Sections: 2. Differential Distributions, CERN-2012-002 (2012), arXiv:1201.3084 [hep-ph]
26. LHC Higgs Cross Section Working Group, Handbook of LHC Higgs Cross Sections: 3. Higgs Properties, CERN-2013-004 (2013), arXiv:1307.1347 [hep-ph]
27. Particle Data Group, Review of particle physics. Phys. Lett. B **204**, 1 (1988)
28. ALEPH, DELPHI, L3, OPAL Collaborations, LEP Working Group for Higgs boson searches, Search for the Standard Model Higgs boson at LEP. Phys. Lett. B **565**, 61 (2003)
29. CDF, D0 Collaborations, Tevatron New Phenomena and Higgs Working Group, Combined CDF and D0 Upper Limits on Standard Model Higgs-Boson Production with up to 6.7 fb^{-1} of Data, FERMILAB-CONF-10-257-E (2010), arXiv:1007.4587 [hep-ex]

30. H. Flacher et al., Revisiting the Global Electroweak Fit of the Standard Model and Beyond with Gfitter, Eur. Phys. J. C **60**, 543 (2009), arXiv:0811.0009 [hep-ph], updated results taken from http://cern.ch/gfitter (Aug 10)
31. J. Ellis, J.R. Espinosa, G.F. Giudice, A. Hoecker, A. Riotto, The probable fate of the standard model. Phys. Lett. B **679**, 369 (2009), arXiv:0906.0954 [hep-ph]
32. ATLAS, CDF, CMS and D0 Collaborations, First combination of Tevatron and LHC measurements of the top-quark mass, ATLAS-CONF-2014-008 (2014), arXiv:1403.4427 [hep-ex]

Chapter 2
Computational Techniques for the LHC

Although the search for the Higgs boson is motivated by the electroweak interaction, a detailed knowledge of quantum chromodynamics (QCD) is required to make precise predictions at a hadron collider such as the LHC. However, these calculations are troublesome; QCD describes the interactions of quarks and gluons, though only the composite hadrons are experimentally observed.

Some key concepts of QCD are introduced in Sect. 2.1, before the simulation of LHC collisions is described in Sect. 2.2. Finally, jets are introduced in Sect. 2.3 as useful tools connecting theoretical calculations with experimental observations.

2.1 Quantum Chromodynamics

QCD is the theory of the strong interaction, describing coloured particles (quarks and gluons, collectively known as partons) [1]. Two crucial features of QCD are *confinement* and *asymptotic freedom*. Confinement refers to the observation that quarks and gluons are only found within colourless hadrons, and never as isolated states. Asymptotic freedom states that, within the hadron, the constituent partons are relatively free to move. Both concepts can be understood in terms of a running coupling constant.

2.1.1 Renormalisation and the Running Coupling Constant

When calculating observables within perturbative quantum field theory, ultraviolet (UV) divergences are often introduced by Feynman diagrams containing loops. Through careful consideration, these UV divergences can be absorbed into renormalised definitions of the coupling constant and particle masses. The idea is that the 'bare' quantities contain compensating divergences, such that the physically measurable quantities are finite:

Fig. 2.1 The running of the strong coupling constant α_S with energy scale Q [6]. Experimental measurements at various scales are also shown

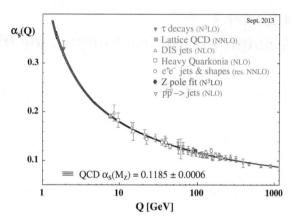

$$g_{\text{physical}} = g_{\text{bare}} + \delta g \quad \text{and} \quad m_{\text{physical}} = m_{\text{bare}} + \delta m \qquad (2.1)$$

where δg and δm are the loop contributions. This procedure is known as *renormalisation*.

It is necessary to introduce an unphysical *renormalisation scale* μ_R, above which loops are absorbed into renormalised quantities, and below which loops are calculated in perturbation theory. The exact details of the separation depend upon the choice of renormalisation scheme [2]. Clearly couplings and masses will depend upon μ_R, though physical observables must not; however, truncation of the perturbative series will result in a residual μ_R dependence. Usually μ_R is chosen to be the energy scale Q of the process under consideration, leading to the concept of a *running coupling constant*.

The QCD coupling constant α_S is shown in Fig. 2.1. At low scales (large distances), α_S is large and the theory is non-perturbative. Though not analytically proven[1], confinement has been verified in this regime by lattice QCD [3]. At high scales (small distances), α_S is small; this is asymptotic freedom [4, 5]. Note that α_{EM} in QED exhibits an opposite trend, though remains perturbative at all accessible energies.

2.1.2 Perturbative QCD

Most interesting LHC processes involve a large momentum transfer, where the partons are asymptotically free. Thus, parton-level cross sections may be calculated with Feynman diagrams as a perturbative series in α_S (which converges since $\alpha_S \ll 1$)

$$\hat{\sigma} = \sum_{m=0}^{\infty} \alpha_S^{k+m} \hat{\sigma}^{(m)} \qquad (2.2)$$

---

[1] A mathematically rigorous proof of confinement is one of seven Millennium Prize Problems of the Clay Mathematics Institute, with a bounty of $1,000,000.

2.1 Quantum Chromodynamics

where the hat denotes a parton-level quantity, k is the number of QCD vertices at tree-level, and $\hat{\sigma}^{(m)}$ is the mth order contribution to the cross section. A *fixed order* calculation truncates the series after n terms, with $n = 0$ being a leading order (LO) calculation, $n = 1$ being a next-to-leading order (NLO) calculation, and so on.

As mentioned in Sect. 2.1.1, the cross section $\hat{\sigma}$ is independent of the renormalisation scale μ_R

$$\frac{d\hat{\sigma}}{d\mu_R} = 0. \tag{2.3}$$

However, real-life calculations always truncate the series after n terms, leaving a residual μ_R dependence. Inserting the truncated series into (2.3), it follows that

$$\frac{d}{d\mu_R} \sum_{m=0}^{n} \alpha_S^{k+m} \hat{\sigma}^{(m)} = -\frac{d}{d\mu_R} \sum_{m=n+1}^{\infty} \alpha_S^{k+m} \hat{\sigma}^{(m)} \tag{2.4}$$

$$= \mathcal{O}\left(\alpha_S^{k+n+1}\right). \tag{2.5}$$

Thus, the residual μ_R dependence can be exploited to probe the effect of missing higher order terms in the series, and estimate the associated uncertainty.

2.1.3 Resummation of Large Logarithms

Fixed order calculations are useful only when the perturbative series is converging, as is usual for an inclusive cross section. However, when considering exclusive observables, there are regions of phase space in which the missing higher order terms contribute as much as the included terms. This often occurs when there is a large separation in the scales of the exclusive observables and the process.

For example, consider the emission of a gluon from an outgoing quark. The scale separation of the hard scatter Q from the soft emission Q_1 introduces Sudakov double logarithmic contributions $\alpha_S^{k+m} L^{2m}$ to the perturbative series, where $L \sim \ln(Q_1/Q)$. Requiring such an emission, the (schematic) structure of the perturbative series becomes

$$\hat{\sigma} \sim \alpha_S^k \left\{ \alpha_S \left(L^2 + L + 1\right) + \alpha_S^2 \left(L^4 + L^3 + L^2 + L + 1\right) + \mathcal{O}\left(\alpha_S^3 L^6\right) \right\}. \tag{2.6}$$

Soft or collinear emissions are defined by $\alpha_S L^2 \approx 1$, such that the logarithms overcome the α_S suppression. Thus, the perturbative nature of the series is spoiled. In (2.6), terms like $\alpha_S^{k+m} L^{2m}$ are called leading logarithms (LLs), terms like $\alpha_S^{k+m} L^{2m-1}$ are called next-to-leading logarithms (NLLs), and so on.

When an observable is sensitive to such large logarithms, they must be *resummed* to all orders in α_S to produce an accurate result. This is usually achieved analytically, but in this example of soft and collinear emissions a *parton shower* Monte Carlo program can be used. This probabilistically generates emissions as it evolves partons from the scale of the hard scatter down to a scale where non-perturbative effects of confinement dominate. This leads to fully-exclusive observables. A parton shower is necessary to produce hadron-level predictions (see Sect. 2.2). Formally they have LL accuracy, though can include many higher order logarithms through the enforcement of physical effects such as energy-momentum conservation and colour coherence.

2.1.4 Parton Distribution Functions

Since confinement binds partons into hadrons, it is the latter that are accelerated and collided at the LHC (in particular protons). Therefore, we need to calculate observables for proton-proton interactions rather than the parton-parton interactions discussed above. Fortunately, the *factorisation theorem* states that the soft non-perturbative physics of the hadron can be treated independently of the hard scatter [7]. Thus, a proton-proton cross section can be formulated as a convolution of the partonic cross section with parton distribution functions (PDFs) of the incoming protons. That is,

$$\sigma(p_1, p_2) = \sum_{a,b} \int_0^1 dx_1 dx_2 f_a\left(x_1, \mu_F^2\right) f_b\left(x_2, \mu_F^2\right)$$
$$\widehat{\sigma}_{ab}\left(x_1 p_1, x_2 p_2, \alpha_S\left(\mu_R^2\right), \frac{Q^2}{\mu_F^2}, \frac{Q^2}{\mu_R^2}\right) \quad (2.7)$$

where f_a is the PDF of parton type a within the proton, p_i is the momentum of proton i, x_i is the momentum fraction of parton i, and Q is the scale of the hard scatter. A sum is performed over all possible parton types (six quark flavours and the gluon).

Echoing renormalisation, factorisation absorbs collinear divergences into universal PDFs which are not *a priori* calculable and must be experimentally constrained. An unphysical *factorisation scale* μ_F is introduced, below which emissions are absorbed into PDFs, and above which they are included in the hard scatter. As with μ_R, truncating the perturbative series introduces a μ_F dependence, which can be exploited to estimate the effect of the missing higher order terms. At LO, $f_a(x, \mu_F)$ is simply the probability of finding a parton of type a with momentum fraction x, when probing the proton at a scale μ_F. However, the interpretation at higher orders is more complicated.

The PDF μ_F scaling is described by the DGLAP equations [8–10]. Thus, an $f_a(x)$ ansatz is made at low μ_F and then experimentally validated at higher scales (e.g with deep inelastic scattering or collider jet data). Figure 2.2 shows some example PDFs.

2.2 Monte Carlo Event Generation

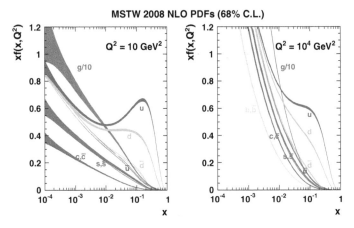

Fig. 2.2 Parton distribution functions fit by the MSTW collaboration, evaluated at $\mu_F^2 = 10\,\text{GeV}^2$ (*left*) and $\mu_F^2 = 10^4\,\text{GeV}^2$ (*right*) [11]. Note that the gluon PDF is suppressed by a factor 10. Reprinted with kind permission from Springer Science and Business Media: European Physical Journal C, **63**, 2009, 189–285, *Parton distributions for the LHC*, A. D. Martin, W. J. Stirling, R. S. Thorne and G. Watt, Fig. 2.1, Copyright 2009. MSTW 2008 NLO PDFs at $Q^2 = 10\,\text{GeV}^2$ and $Q^2 = 10^4\,\text{GeV}^2$

2.2 Monte Carlo Event Generation

Monte Carlo (MC) event generators provide a fully-exclusive hadron-level simulation of pp collision events at the LHC [12]. This section will describe the basic features of a simulated event, before discussing some more advanced techniques that shall be used throughout the thesis.

2.2.1 The Anatomy of an Event

Figure 2.3 shows how the MC event generation is factorised into several components, each describing a certain regime of momentum transfer.

Hard scatter
   The high scale process can be selected as desired (e.g Higgs boson production via gluon-gluon fusion). The relevant parton-level matrix elements (MEs) are calculated using fixed order perturbative QCD, either by the event generator itself or an external program. Historically, these MEs were usually LO, though improvements are discussed in Sects. 2.2.3 and 2.2.4.

Parton distribution functions (PDFs)
   Incoming parton momenta are sampled from a proton PDF, usually probed at the scale of the hard scatter ($\mu_F = Q$). The LHAPDF interface [14] provides access to the PDFs of several fitting collaborations, such as CTEQ [15],

Fig. 2.3 Schematic diagram of a simulated ttH event, showing how factorisation allows the physics at different scales of momentum transfer Q to be treated independently [13]. At high-Q is the hard scatter (*red circle*). As the scale evolves down, partons are radiated in the initial state (*blue*) and final state (*red*). At low-Q, incoming partons are confined to the beam protons, while outgoing partons hadronise (*green blobs*). The underlying event comprises multiple partonic interactions (*purple blob*) and beam remnants (*blue blobs*). Photons and leptons (*yellow*) are also radiated

MSTW [11] and NNPDF [16]. PDFs differ because they are fit with different subsets of experimental data, massive quark treatments, parametrisation models and $\alpha_S(m_Z)$ values.

Final state radiation (FSR)

Soft and collinear radiation from outgoing partons is simulated by a universal parton shower, evolving the scale from the hard scatter to the hadronisation scale of ~ 1 GeV. The successive emissions are ordered to avoid double-counting—typical order parameters are virtuality, transverse momentum and opening angle.

2.2 Monte Carlo Event Generation

For the correct treatment of soft emissions, it is vital to preserve colour coherence. This is inherent in an angular ordered shower, but must be manually implemented otherwise. Alternatively, a *dipole shower* considers emissions from colour-connected pairs of partons, and is also inherently coherent.

Initial state radiation (ISR)

Soft and collinear radiation from incoming partons is similarly described by a parton shower. However, the small probability of evolving two partons with the kinematics required by the hard process necessitates a *backwards evolution*. Thus, the probability that a parton originated from one of higher momentum and lower scale is calculated, rather than an emission probability.

Hadronisation

The confinement of partons to hadrons is non-perturbative, and must be described by a hadronisation model. The *string model* stretches strings between colour partners. At some distance it becomes favourable to convert the potential energy to a $q\bar{q}$ pair, breaking the string. Once there is insufficient energy to create $q\bar{q}$ pairs, the hadrons 'freeze out'. The *cluster model* splits gluons into $q\bar{q}$ pairs, which group into colourless clusters with a mass spectrum predicted by QCD. These clusters then decay to the physical hadrons. Note that all hadronisation models require tuning to experimental data.

Hadron and τ decays

Many of the hadrons produced during hadronisation are unstable, and must be decayed to particles that are stable on a detector traversal timescale, while observing conservation laws and measured branching ratios. Similarly, the τ lepton must be decayed, hadronically or leptonically.

Multiple partonic interactions (MPI)

The *underlying event* (UE) is the additional soft hadronic activity caused by partons inactive in the hard scatter. It comprises the breakup of the beam remnants and *multiple partonic interactions* (MPI) between the protons. The size of the MPI activity is correlated to the scale of the hard scatter.

In order to calculate the number of additional interactions, the spatial distribution of partons within the proton must be modelled, the impact parameter of the *pp* collision must be known, and an IR cut-off must be imposed. This requires non-perturbative models that must be tuned to experimental data.

QED radiation

Electrically charged particles can emit photons at any stage of the event generation.

2.2.2 Summary of Event Generators

Three event generators are commonly used at the LHC, mainly differing in their choice of hadronisation and MPI models, and their parton shower order parameter. Efforts to rewrite the older Fortran-based programs in C++ have led to a generation of 'out-of-date' Fortran programs that are no longer actively developed. Even so, they are still in common usage, and so are included in the descriptions below.

Herwig

> HERWIG (Fortran) [17] and HERWIG++ (C++) [18] both employ an angular ordered parton shower and a cluster hadronisation model. An MPI model is included in HERWIG++, but in HERWIG this was provided by JIMMY [19].

Pythia

> PYTHIA 6 (Fortran) [20] and PYTHIA 8 (C++) [21] both use a string hadronisation model and an advanced MPI model. PYTHIA 8 uses a dipole shower ordered in transverse momentum, whereas PYTHIA 6 offers a choice of virtuality or transverse momentum ordered parton showers with colour coherence implemented manually.

Sherpa

> SHERPA (C++) [22] uses a dipole shower ordered in transverse momentum, which is convenient for multi-leg merging (see Sect. 2.2.3). It uses a cluster hadronisation model and an MPI model similar to that of PYTHIA 8.

2.2.3 Multi-leg Merging

Although a parton shower excellently describes the emissions of large numbers of soft and collinear partons, it fails to accurately model hard and isolated emissions. It can be desirable to describe these using fixed order MEs, which are better suited to the task. In doing so, a couple of immediate issues arise. First, we require a smooth transition from the emissions of an ME to those of the parton shower. Second, each ME is inclusive, and attempting to combine MEs of differing multiplicity naturally leads to problems of double counting.

By using a merging prescription, such as the CKKW-L algorithm [23, 24] employed by SHERPA or the MLM algorithm [25] employed by ALPGEN [26] and MADGRAPH [27], it is possible to consistently combine LO matrix elements with differing multiplicities, whilst matching to the parton shower correctly. This does require the introduction of a merging scale though. This scale separates the ME and parton shower descriptions of the emissions, though the details of the separation depend upon the merging prescription.

2.2 Monte Carlo Event Generation

2.2.4 NLO Matching

It is also possible to match an NLO ME to a parton shower, to improve the accuracy of both the normalisation and distribution of observables [28]. Such a calculation must include the LO, virtual-loop and real-emission diagrams, while mapping smoothly onto the parton shower for soft emissions. There are currently two valid matching prescriptions:

MC@NLO
Simply adding a parton shower to an NLO ME introduces double counting of emissions. The MC@NLO method compensates for this overlap through a correction to the NLO calculation. This correction renders the ME dependent upon the parton shower used in the MC event generator, and is also a source of negatively weighted events.

Originally implemented in the MC@NLO program for matching to HERWIG [29] and HERWIG++ [30], the method has now been automated within the AMC@NLO program [31] and extended for use with PYTHIA 6 and PYTHIA 8 [32]. It is now also included in SHERPA.

POWHEG
The POWHEG method requires the hardest emission to always be generated by the ME. It achieves the correct hard and soft behaviour by convolving the LO ME with a modified Sudakov factor, and then reweighting the differential cross section to the NLO result. Thus the ME is independent of the subsequent parton shower. However, if the parton shower is not transverse momentum ordered, it is necessary to use truncated and vetoed parton showers to correctly fill the phase space.

Originally implemented in POWHEGBOX [33–35], variants are now also included in HERWIG++ and SHERPA.

2.2.5 Additional Considerations

Detector simulation
In order to compare MC events to experimental events recorded at the LHC, it is vital to simulate how the outgoing particles interact with the detector. This is also necessary to calibrate the detector response and estimate efficiencies. GEANT4 [36, 37] is used to simulate the energy deposition of each particle during its trajectory through the ATLAS detector (see Chap. 3). Since long-lived particles will decay *en route*, particles with lifetime $c\tau > 10$ mm are decayed by GEANT4 rather than the MC generator. The majority of the simulation time is spent modelling the complex calorimeter geometry; in some cases

the simulation is performed by ATLFAST-II [38], which contains a simplified calorimeter simulation.

Digitisation converts the energy deposition into readout voltages and currents. Following this, the events can be treated like experimental collision events.

Pile-up simulation

As described in Chap. 3, each LHC bunch crossing can result in soft proton-proton interactions, known as *pile-up*, in addition to the hard process. This obscures the interesting physics and is important to model accurately.

In-time pile-up (same bunch crossing as the hard process) is modelled by overlaying simulated energy deposits from soft pp interactions generated with PYTHIA 8. The number of overlaid events depends upon the beam conditions (see Sect. 3.2).

Out-of-time pile-up (different bunch crossing to the hard process) affects detector sub-systems whose latency is longer than the bunch spacing. For such sub-systems, signals from out-of-time pile-up are overlaid with corresponding time shifts; again this depends on the beam conditions.

2.2.6 Parton Shower Tuning Study

When studying MC modelling uncertainties in the ggF process (see Sect. 5.2), a discrepancy was observed at high jet multiplicity between POWHEGBOX+PYTHIA 8 and POWHEGBOX+PYTHIA 6 (see the green and black lines in Fig. 2.4). It is also observed in other electroweak processes.

The hadronisation and UE models of standalone PYTHIA 8 have been tuned to ATLAS UE data with a variety of PDF sets (known as AU2 tunes) [39]. However, the parton shower was not tuned since the default settings successfully described experimental data.

When modelling ggF with POWHEGBOX, the AU2-CT10 tune was used in order to match the PDFs used in the matrix element calculation. Technically speaking, a dedicated POWHEGBOX+PYTHIA 8 tune should have been used, but this was unavailable. Unfortunately, a couple of issues had a negative impact on the NLO-PS matching. First, the parton shower evolves α_S at LO, whilst NLO PDFs were used in the shower. Second, there was a mismatch between the α_S used in POWHEGBOX, $\alpha_S(m_Z) = 0.118$, and the default value in the parton shower, $\alpha_S(m_Z) = 0.137$. The effect of these issues is shown in Fig. 2.4.

Identification of this poor matching has led to improvements in the latest round of MC tuning, where dedicated POWHEGBOX+PYTHIA 8 tunes are fit using an adjusted parton shower [40]. These shall be used in Run II.

2.3 Jet Algorithms

We have seen in Sect. 2.2 how coloured partons produced in a hard subprocess (in the ME) or radiated from the incoming partons (ISR) will each produce a shower of partons, which subsequently hadronise. By measuring the energy and direction of the resulting collimated *jet* of hadrons, it is possible to infer information about the original quark or gluon. This is very useful for probing the perturbative hard scatter, whilst remaining fairly insensitive to poorly understood hadronisation effects.

A *jet algorithm* defines how the large number of final state particle four-momenta are grouped into a small number of jet four-momenta. Such an algorithm should satisfy a number of criteria, the most important being infrared and collinear safety [41]. This requires that the jets are insensitive to additional soft or collinear emissions.

Multiple jet algorithms are implemented in the FASTJET software library [42]. In particular, *sequential recombination algorithms* are popular at the LHC, which iteratively combine the closest pair of particles according to some distance measure d_{ij}.

Consider an algorithm where all the inter-particle distances d_{ij} and particle-beam distances d_{iB} are calculated. If the minimum of these is a d_{ij} rather than a d_{iB}, then particles i and j are combined into single new particle. If the minimum is a d_{iB}, then particle i is declared a jet and removed from the list of particles. Then the algorithm restarts. We define the distances

$$d_{ij} = \min\left(p_{Ti}^{2m}, p_{Tj}^{2m}\right)\frac{\Delta R_{ij}^2}{R^2}, \qquad \Delta R_{ij}^2 = (y_i - y_j)^2 + (\phi_i - \phi_j)^2 \quad (2.8)$$

$$d_{iB} = p_{Ti}^{2m} \quad (2.9)$$

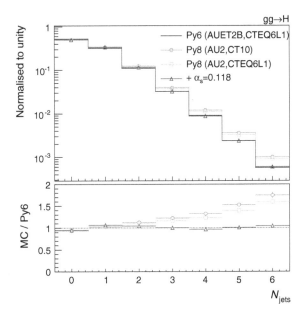

Fig. 2.4 Jet multiplicity produced by POWHEGBOX+PYTHIA 8 with a selection of shower tunes. The *green circles* correspond to the tune used in the analysis. The *red squares* change the parton shower PDFs from CT10 to CTEQ6L1. The *blue triangles* additionally change the parton shower $\alpha_S(m_Z)$ from 0.137 to 0.118 (as used in POWHEGBOX). POWHEGBOX+PYTHIA 6 is shown in *black* for reference, and is in good agreement with POWHEGBOX+HERWIG (not shown)

where p_{Ti}, y_i and ϕ_i are the transverse momentum, rapidity and azimuthal angle of particle i with respect to the beam axis, respectively. R and m are parameters of the algorithm, with R effectively determining the size of the jet.

With $m = 1$, known as the k_T algorithm, combinations between soft particles are favoured. This follows the evolution of QCD, but leads to rather irregular jet shapes.

With $m = -1$, known as the anti-k_T algorithm [43], combinations between hard particles are favoured. This means the jets grow outwards from a hard 'seed', ultimately producing more circular jets. However, the jet substructure can no longer be used to infer details of the jet evolution history. The jets used in this thesis were anti-k_T jets with $R = 0.4$, and their reconstruction shall be described in detail in Sect. 4.2.5.

References

1. R.K. Ellis, W.J. Stirling, B.R. Webber, *QCD and Collider Physics* (Cambridge University Press, Cambridge, 1996)
2. I.J.R. Aitchison, A.J.G. Hey, *Gauge Theories in Particle Physics*, 3rd edn. (Taylor and Francis, Abingdon, 2003)
3. K.G. Wilson, Confinement of quarks. Phys. Rev. D **10**, 2445 (1974)
4. D.J. Gross, F. Wilczek, Ultraviolet behavior of non-Abelian Gauge theories. Phys. Rev. Lett. **30**, 1343 (1973)
5. H.D. Politzer, Reliable perturbative results for strong interactions? Phys. Rev. Lett. **30**, 1346 (1973)
6. Particle Data Group, Review of particle physics. Phys. Rev. D **86**, 010001 (2012), and 2013 partial update for the 2014 ed
7. J.C. Collins, D.E. Soper, Parton distribution and decay functions. Nucl. Phys. B **194**, 445 (1982)
8. V.N. Gribov, L.N. Lipatov, Deep inelastic ep scattering in perturbation theory. Sov. J. Nucl. Phys. **15**, 438 (1972)
9. G. Altarelli, G. Parisi, Asymptotic freedom in parton language. Nucl. Phys. B **126**, 298 (1977)
10. Y. L. Dokshitser, Calculation of structure functions of deep inelastic scattering and e^+e^- annihilation by perturbative theory in quantum chromodynamics. Sov. Phys.-JETP **46**, 641 (1977)
11. A.D. Martin, W.J. Stirling, R.S. Thorne, G. Watt, Parton distributions for the LHC. Eur. Phys. J. C **63**, 189 (2009), arXiv:0901.0002 [hep-ph]
12. A. Buckley et al., General-purpose event generators for LHC physics. Phys. Rep. **504**, 145 (2011), arXiv:1101.2599 [hep-ph]
13. S. Höche, Matching to matrix elements, in MCnet-LPCC Summer School on Monte Carlo Event Generators for LHC, Geneva (2012)
14. M.R. Whalley, D. Bourilkov, R.C. Group, The Les Houches accord PDFs (LHAPDF) and LHAGLUE, in HERA and the LHC, Hamburg (2005), arXiv:hep-ph/0508110
15. H.-L. Lai et al., New parton distributions for collider physics. Phys. Rev. D **82**, 074024 (2010), arXiv:1007.2241 [hep-ph]
16. R.D. Ball et al., Parton distributions with LHC data. Nucl. Phys. B **867**, 244 (2013), arXiv:1207.1303 [hep-ph]
17. G. Corcella et al., HERWIG 6: an event generator for hadron emission reactions with interfering gluons (including supersymmetric processes). JHEP **0101**, 010 (2001), arXiv:hep-ph/0210213
18. M. Bähr et al., Herwig++ physics and manual. Eur. Phys. J. C **58**, 639 (2008), arXiv:0803.0883 [hep-ph]

References

19. J.M. Butterworth, J.R. Forshaw, M.H. Seymour, Multiparton interactions in photoproduction at HERA. Z. Phys. C **72**, 637 (1996), arXiv:hep-ph/9601371
20. T. Sjöstrand, S. Mrenna, P. Skands, PYTHIA 6.4 physics and manual. JHEP **0605**, 026 (2006), arXiv:hep-ph/0603175
21. T. Sjöstrand, S. Mrenna, P. Skands, A brief introduction to PYTHIA 8.1. Comput. Phys. Commun. **178**, 852 (2008), arXiv:0710.3820[hep-ph]
22. T. Gleisberg et al., Event generation with SHERPA 1.1. JHEP **0902**, 007 (2009), arXiv:0811.4622 [hep-ph]
23. S. Catani, F. Krauss, R. Kuhn, B.R. Webber, QCD matrix elements + parton showers. JHEP **0111**, 063 (2001), arXiv:hep-ph/0109231
24. L. Lonnblad, Correcting the color-dipole cascade model with fixed order matrix elements. JHEP **0205**, 046 (2002), arXiv:hep-ph/0112284
25. S. Hoeche et al., Matching parton showers and matrix elements (2006), arXiv:hep-ph/0602031
26. M.L. Mangano, M. Moretti, F. Piccinini, R. Pittau, A.D. Polosa, ALPGEN, a generator for hard multiparton processes in hadronic collisions. JHEP **0307**, 001 (2003), arXiv:hep-ph/0206293
27. J. Alwall, M. Herquet, F. Maltoni, O. Mattelaer, T. Stelzer, MadGraph 5: going beyond. JHEP **1106**, 128 (2011), arXiv:1106.0522 [hep-ph]
28. P. Nason, B. Webber, Next-to-leading-order event generators. Ann. Rev. Nucl. Part. Sci. **62**, 187 (2012), arXiv:1202.1251 [hep-ph]
29. S. Frixione, B.R. Webber, Matching NLO QCD computations and parton shower simulations. JHEP **0206**, 029 (2002), arXiv:hep-ph/0204244
30. S. Frixione, F. Stoeckli, P. Torrielli, B.R. Webber, NLO QCD corrections in Herwig++ with MC@NLO. JHEP **1101**, 053 (2011), arXiv:1010.0568 [hep-ph]
31. V. Hirschi et al., Automation of one-loop QCD corrections. JHEP **1105**, 044 (2011), arXiv:1103.0621 [hep-ph]
32. P. Torrielli, S. Frixione, Matching NLO QCD computations with PYTHIA using MC@NLO. JHEP **1004**, 110 (2010), arXiv:1002.4293 [hep-ph]
33. P. Nason, A new method for combining NLO QCD with shower Monte Carlo algorithms. JHEP **0411**, 040 (2004), arXiv:hep-ph/0409146
34. S. Frixione, P. Nason, C. Oleari, Matching NLO QCD computations with parton shower simulations: the POWHEG method. JHEP **0711**, 070 (2007), arXiv:0709.2092 [hep-ph]
35. S. Alioli, P. Nason, C. Oleari, E. Re, A general framework for implementing NLO calculations in shower Monte Carlo programs: the POWHEG BOX. JHEP **1006**, 043 (2010), arXiv:1002.2581 [hep-ph]
36. S. Agostinelli et al., Geant4—a simulation toolkit. Nucl. Instrum. Methods **A506**, 250 (2003)
37. ATLAS Collaboration, The ATLAS simulation infrastructure, Eur. Phys. J. C **70**, 823 (2010), arXiv:1005.4568[physics.ins-det]
38. ATLAS Collaboration, The simulation principle and performance of the ATLAS fast calorimeter simulation FastCaloSim, ATL-PHYS-PUB-2010-013 (2010)
39. ATLAS Collaboration, Summary of ATLAS Pythia 8 tunes. ATL-PHYS-PUB-2012-003 (2012)
40. ATLAS Collaboration, Example ATLAS tunes of Pythia8, Pythia6 and Powheg to an observable sensitive to Z boson transverse momentum. ATL-PHYS-PUB-2013-017 (2013)
41. G.P. Salam, Towards jetography. Eur. Phys. J. C **67**, 637 (2010), arXiv:0906.1833 [hep-ph]
42. M. Cacciari, G.P. Salam, G. Soyez, FastJet user manual. Eur. Phys. J. C **72**, 1896 (2012), arXiv:1111.6097 [hep-ph]
43. M. Cacciari, G.P. Salam, G. Soyez, The anti-k_t jet clustering algorithm. JHEP **0804**, 063 (2008), arXiv:0802.1189 [hep-ph]

Chapter 3
The ATLAS Experiment

The ATLAS experiment is a general-purpose particle detector operated by an international collaboration of more than 3000 scientists. It is designed to search for a broad range of new phenomena by precisely studying the collisions of high energy protons, which are provided by the Large Hadron Collider (LHC) accelerator at CERN, Geneva.

The LHC and the *pp* collision data it provides are described in Sect. 3.1 and Sect. 3.2 respectively. Then, the ATLAS detector is described in Sect. 3.3.

3.1 The Large Hadron Collider

The LHC is the world's largest and most energetic particle accelerator. It is installed in a 27 km tunnel at a mean depth of 100 m beneath the French–Swiss border, which was previously occupied by the Large Electron–Positron collider (LEP). It accelerates beams of protons (or lead ions) to high energy and then collides them within the ATLAS, CMS, LHCb and ALICE detectors. Although the design centre-of-mass (CM) collision energy is 14 TeV, it operated at 7 and 8 TeV during Run I, as described in Sect. 3.2.

The LHC proton beams have humble beginnings as hydrogen molecules in a standard gas bottle, before being stripped of their electrons and entering the LHC accelerator complex shown in Fig. 3.1. The LHC accelerator features 16 radio-frequency cavities to provide acceleration and restore energy losses, 1232 dipole magnets to bend the beam into a nearly circular path, 392 quadrupole magnets to focus the beam, and many other magnetic components.[1] Most of these magnets rely on unprecedented superconducting twin-bore magnet technology operating at cryogenic temperatures.

For experiments to be sensitive to rare processes with small cross sections, such as Higgs boson production, the detectors must record a large integrated luminosity.

---

[1]Former DG of CERN Christopher Llewellyn Smith chose the magnet colour to resemble Oxford blue.

Fig. 3.1 The LHC accelerator complex at CERN. The successively higher energy accelerators are: Linear Accelerator 2 (LINAC 2), Proton Synchrotron Booster, Proton Synchrotron (PS), Super Proton Synchrotron (SPS), and finally the Large Hadron Collider (LHC). The four main LHC detectors are also shown

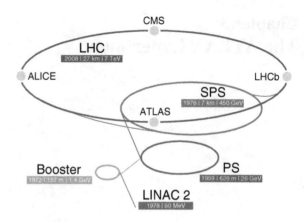

This is seen by considering the expected number of events produced for process i

$$N_i = \sigma_i \int L \, \mathrm{dt} \tag{3.1}$$

where σ_i is the cross section and L is the instantaneous luminosity, a figure of merit for a collider. The instantaneous luminosity can be increased by optimising the parameters of

$$L = \frac{N_b^2 n_b f_{\mathrm{rev}}}{4\pi \Sigma_x \Sigma_y} F \tag{3.2}$$

where N_b is the number of particles per bunch, n_b is the number of bunches per beam, f_{rev} is the revolution frequency, Σ_x and Σ_y are the x and y components of the beam size, and F is a reduction factor due to the crossing angle at the interaction point. The LHC is designed to hold 2808 proton bunches, corresponding to 25 ns bunch spacing, each containing 1.15×10^{11} protons. The other design parameters are $\Sigma_{x,y} = 16.7$ μm and $f_{\mathrm{rev}} = 11.25$ kHz, giving a design luminosity of $L \approx 10^{34}$ cm^{-2} s^{-1} [1].

A trade-off for higher luminosity is a larger number of additional proton-proton interactions, known as *pile-up*. Although a rare interesting event will trigger the detector readout, these common uninteresting events will simultaneously be recorded, obscuring the interesting physics and degrading detector performance. Increasing N_b gives more interactions within the same bunch crossing, known as *in-time pile-up*. For large n_b, the bunch spacing can be shorter than the detector latency, and interactions from other bunch crossings can affect the measurement; this is known as *out-of-time pile-up*. Reducing $\Sigma_{x,y}$ will increase both types of pile-up.

3.2 pp Collision Data

It is clear from (3.1) that the luminosity delivered to the detector is a key input when studying pp collisions at the LHC. It is directly proportional to the expected number of events, and uncertainties in its value will be propagated to measured cross sections. The measurement of the luminosity delivered to the ATLAS detector is described in Sect. 3.2.1, followed by a description of the dataset used in this thesis.

3.2.1 Luminosity Measurement

Beam losses incurred by the collisions cause the luminosity to decay (a typical run lasts ≈ 10 hours). Thus, it is necessary to measure the instantaneous luminosity in real-time.

At the LHC, the number of inelastic pp interactions per bunch crossing follows a Poisson distribution, with a mean value μ. As mentioned in Sect. 3.1, a large luminosity results in $\mu > 1$ (a condition known as pile-up). Thus, the luminosity L can be monitored "online" by measuring the observed number of interactions per crossing μ_{vis}, using [2]

$$L = \frac{\mu n_b f_{\text{rev}}}{\sigma_{\text{inel}}} = \frac{\mu_{\text{vis}} n_b f_{\text{rev}}}{\sigma_{\text{vis}}} \qquad (3.3)$$

where σ_{inel} is the inelastic pp cross section. The expression is rewritten with "visible" quantities, owing to inefficiencies in the detector and algorithm used to measure μ.

The beam conditions monitor (BCM) and LUCID detectors, respectively situated 2 and 17 m down the beamline, each count the number of activated readout channels per bunch crossing, which is highly correlated with μ_{vis}. The BCM consists of 16 small diamond sensors, and was primarily designed to issue beam-abort requests when beam losses risk damaging the ATLAS detector. LUCID comprises 16 tubes of C_4F_{10} gas, which radiate and collect Cherenkov photons when struck by charged particles.

BCM and LUCID are calibrated during dedicated van der Meer (vdM) scans, effectively determining σ_{vis} in (3.3). In a vdM scan, event rates are measured while the beams are separated in steps of known distance, allowing direct measurement of beam sizes Σ_x and Σ_y. The absolute luminosity is then determined through (3.2). The uncertainty in the vdM calibration dominates the uncertainty in the delivered luminosity.

Additional methods, such as measuring average particle rates with the ATLAS calorimeters, can be used to improve the luminosity estimation offline.

Table 3.1 Summary of *pp* collision data during LHC Run I

	2010	2011	2012	Design
Centre-of-mass energy (TeV)	7	7	8	14
Minimum bunch spacing (ns)	150	50	50	25
Peak luminosity (10^{33} cm^{-2} s^{-1})	0.2	3.6	7.7	10
Delivered luminosity (fb^{-1})	0.047	5.46	22.8	–
Recorded luminosity (fb^{-1})	0.044	5.08	21.3	–
Luminosity uncertainty $\delta L/L$	3.5 %	1.8 %	2.8 %	–

Luminosities use the offline calibration

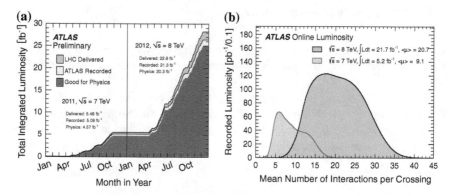

Fig. 3.2 **a** Cumulative luminosity delivered (*green*), recorded (*yellow*), and declared 'good for physics' (*blue*) during 2011 and 2012. Luminosities use the offline calibration. **b** Mean number of interactions per bunch crossing μ for the 2011 (*blue*) and 2012 (*green*) datasets, calculated with an inelastic *pp* cross section of 71.5 mb at $\sqrt{s} = 7$ TeV and 73.0 mb at $\sqrt{s} = 8$ TeV. Luminosities use the online calibration

3.2.2 Run I dataset

Data-taking operations during Run I of the LHC were incredibly successful, and some important parameters of the *pp* datasets are summarised in Table 3.1 and Fig. 3.2. These show that a larger dataset was obtained in 2012 compared with 2011, but at the expense of a higher pile-up environment.

3.3 The ATLAS Detector

ATLAS is a general-purpose particle detector for probing hadron-hadron collisions [3], from precise measurements of Standard Model (SM) processes to searches for signatures of new physics. As such, it must achieve a good performance in the reconstruction of all physics objects interacting with the detector (leptons, photons and jets), and infer the existence of non-interacting particles through transverse momentum imbalance. A good vertex resolution is needed for jet flavour tagging

3.3 The ATLAS Detector

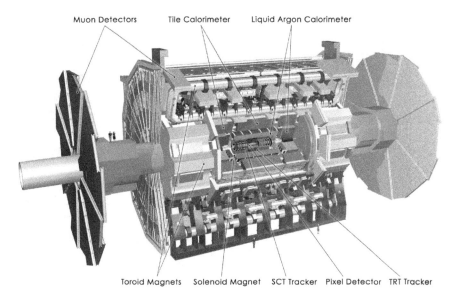

Fig. 3.3 Cut-away view of the ATLAS detector [3]. It has a height of 25 m, a length of 44 m and weighs 7000 tonnes

and to distinguish pile-up events. Other requirements include a fast trigger system to select interesting events to record.

The design of ATLAS exhibits cylindrical and forward-backward symmetries with the nominal interaction point at the centre.[2] The three detector sub-systems (tracking, calorimetry and the muon spectrometer) follow a consistent design of a central barrel with end-caps at both ends, giving a high degree of hermeticity (see Fig. 3.3). Barrel components are arranged on concentric cylinders around the beam axis, while end-cap components are on disks perpendicular to the beam axis.

Figure 3.4 shows how different particles interact with the various parts of the detector. Next to the beam pipe is the inner detector which precisely tracks the trajectories of charged particles. Then there are the electromagnetic and hadronic calorimeters which absorb and measure the energy of interacting particles. Since high-p_T muons act as minimum ionising particles, they survive the calorimeters and are tracked by the muon spectrometer. Superconducting solenoid and toroid magnets provide magnetic fields to the inner detector and muon spectrometer respectively, allowing the momenta of charged particles to be measured from the curvature of their tracks.

---

[2]ATLAS uses a right-handed coordinate system with its origin at the nominal interaction point. The x-axis points to the centre of the LHC ring, the y-axis points upwards, and the z-axis points along the beam line. Positions and directions within the detector are given in spherical coordinates (r, θ, ϕ) where r is the radial distance, θ is the polar angle and ϕ is the azimuthal angle. Usually the polar angle is replaced with pseudorapidity $\eta = -\ln \tan(\theta/2)$. The distance between two positions in $\eta - \phi$ space is $\Delta R = \sqrt{\Delta \eta^2 + \Delta \phi^2}$. Observables labelled "transverse" are projected into the $x - y$ plane.

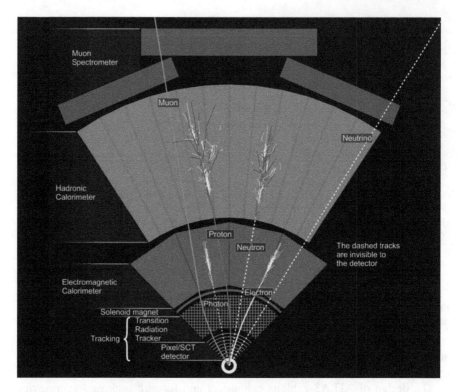

Fig. 3.4 Cross-sectional view of ATLAS, showing how different particles interact with the sub-detectors. ATLAS Experiment © 2013 CERN

3.3.1 Tracking

Pattern recognition algorithms are used to precisely track the trajectories of charged particles through the inner detector (ID), shown in Fig. 3.5. The ID is immersed in a 2 T solenoidal magnetic field, which enables the momentum of a particle to be measured from the curvature of its track. Momentum and vertex measurements both require an excellent spatial resolution, which is achieved through fine detector granularity. The ID covers the region $|\eta| < 2.5$ and consists of three complementary sub-detectors:

Pixel detector
   The pixel detector is installed in three layers closest to the beam pipe, and therefore requires the highest granularity to handle the large particle fluxes. It comprises more than 80 million silicon pixels, each of area $50 \times 400\ \mu m^2$ and thickness 250 μm. The intrinsic accuracy is 10 μm × 115 μm in $r\phi - z$ ($r\phi - r$) space in the barrel (end-cap).

3.3 The ATLAS Detector

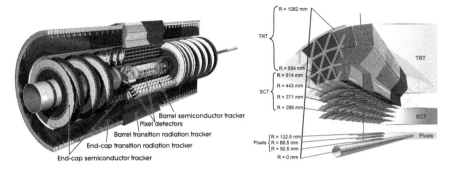

Fig. 3.5 The ATLAS inner detector: cut-away view (*left*) and cross-sectional view showing a trajectory through the three sub-detectors (*right*) [3]. The inner detector has a length of 6.2 m and a radius of 1.1 m. The beam pipe radius is 29 mm

A silicon particle detector consists of a reverse-biased p-n junction. When a charged particle passes through the depletion region it creates an electron-hole pair, which travel to the respective electrodes and produce a signal current.

Semiconductor tracker (SCT)

The SCT features 15,912 silicon strip sensors, each consisting of 770 strips with a pitch of 80 μm and a length of 6 cm. Pairs of sensors are sandwiched together into modules with a stereo angle of 40 mrad, enabling the coordinate parallel to the strip to be measured. There are four (nine) layers of modules in the barrel (end-cap), which ensures that each track passes through at least four modules. The intrinsic accuracy is 17 μm × 580 μm in $r\phi - z$ ($r\phi - r$) space for the barrel (end-cap).

Transition radiation tracker (TRT)

The TRT features 370,000 drift chambers, known as straws, which simultaneously function as a straw tracker to enhance particle tracking and as a transition radiation detector to aid electron identification. Straws are aligned with the beam pipe in the barrel and radially in the end-caps. The TRT covers the region $|\eta| < 2.0$.

Each 4 mm diameter straw has a Kapton® wall with a conductive coating, which acts as a cathode at −1530 V, and a central tungsten wire anode. The straws are filled with a xenon-based gas mixture that is ionised by a traversing charged particle. The freed electrons drift to the anode and produce a signal current. The drift time is used to measure the impact parameter of the incident charged particle relative to the anode and, since a track typically traverses 36 straws, collectively this information yields an intrinsic accuracy of 130 μm in $r\phi$ (much smaller than the straw diameter). However, there is no tracking information parallel to the straw.

The layers of straws are interleaved with polypropylene radiator fibres or foils, and the changes in refractive index cause charged particles to emit X-ray transition radiation (TR). The TR is absorbed by the xenon gas in the straws;

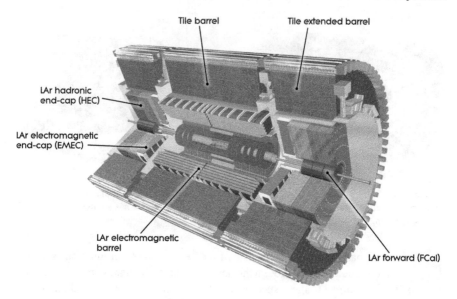

Fig. 3.6 Cut-away view of the ATLAS calorimeter system [3]

TR signals are distinguished from tracking signals by a higher threshold. Since the probability of TR is proportional to the particle's γ-factor, for a given energy lighter particles produce more TR than heavier particles. This aids electron-pion discrimination.

3.3.2 Calorimetry

Particle energies are measured with sampling calorimeters, which consist of alternating layers of absorber and active material. The dense absorber causes energy deposition via a particle shower, whilst the active material produces a signal proportional to the sampled energy. The length scale of energy loss is the radiation length X_0 for electrons and photons[3] and the nuclear interaction length λ for hadronic showers,[4] while muons act as minimum ionising particles and escape the calorimeters. Since punch-through into the muon system must be minimised, X_0 and λ set the size of the calorimeter. Conversely, limits to detector size constrain the materials that can be used.

---

[3] High-energy electrons (photons) mostly lose energy in matter by bremsstrahlung (e^+e^- pair production). In this regime, X_0 is (a) the mean distance in which an electron loses all but e^{-1} of its energy, (b) $\frac{7}{9}$ of the mean free path of a photon, and (c) the characteristic scale of electromagnetic showers.

[4] Hadrons lose energy in matter through inelastic hadronic interactions, forming hadronic showers (though neutral pions create electromagnetic showers). λ is (a) the mean free path of a hadron, and (b) the characteristic scale of hadronic showers.

3.3 The ATLAS Detector

The ATLAS calorimeter consists of an inner electromagnetic calorimeter (for electrons and photons) and an outer hadronic calorimeter (for jets), as shown in Fig. 3.6. They cover the regions $|\eta| < 3.2$ and $|\eta| < 4.9$ respectively. The total thickness of the electromagnetic calorimeter corresponds to more than $22X_0$ and the total thickness of the entire calorimeter corresponds to more than 10λ.

Electromagnetic calorimeter (ECal)

The ECal uses a lead absorber and a liquid argon (LAr) active material arranged in an accordion geometry, which ensures uniform ϕ-coverage. The LAr is ionised by charged particles, and the freed electrons produce a signal current in the readout electrodes. LAr was chosen for its linear behaviour, stability and radiation-hardness.

A fine granularity is required to make precise electron and photon measurements, and to distinguish single photons from $\pi^0 \to \gamma\gamma$. Thus, the ECal is typically segmented into three layers with a minimum $\Delta\eta \times \Delta\phi$ granularity of 0.003×0.025. The transition between the barrel and the end-cap, $1.37 < |\eta| < 1.52$, is used for detector services. Since this material is difficult to model, this "crack" region is usually excluded when reconstructing electrons.

For $|\eta| < 1.8$, a LAr presampler is installed before the first lead layer. This enables estimation of the energy lost by electrons and photons before encountering the ECal.

Hadronic calorimeter (HCal)

The HCal consists of a tile calorimeter covering $|\eta| < 1.7$, a hadronic end-cap (HEC) covering $1.5 < |\eta| < 3.2$, and a forward calorimeter (FCal) covering $3.1 < |\eta| < 4.9$. The HCal has a coarser granularity than the ECal: it is typically segmented into three layers with a minimum $\Delta\eta \times \Delta\phi$ granularity of 0.1×0.1.

The tile calorimeter uses a steel absorber with scintillator tiles as the active material. The scintillation light is collected by optical fibres and turned into a signal by photomultiplier tubes. The HEC and FCal both use LAr active material. The HEC uses a copper absorber, whereas the FCal uses copper in the first layer (for electromagnetic measurements) and tungsten in the two subsequent layers.

3.3.3 Muon Spectrometer

The muon spectrometer (MS) provides precise tracking of muons that have exited the calorimeters. A huge air-core toroid magnet system generates a 0.5 T (1 T) field in the barrel (end-caps), enabling momentum to be inferred from track curvature measurements. Four types of tracking chamber are installed in three layers (see Fig. 3.7):

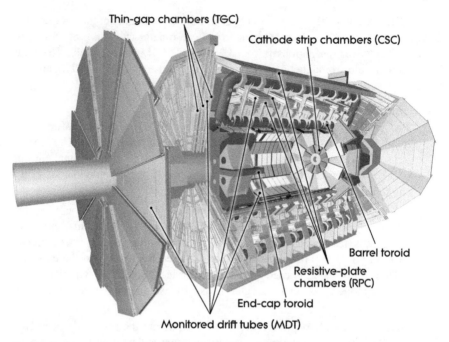

Fig. 3.7 Cut-away view of the ATLAS muon spectrometer [3]

Monitored drift tubes (MDTs) and cathode strip chambers (CSCs)
MDTs and CSCs provide precise momentum measurement, with a resolution of about 40 μm in the bending plane. MDTs cover the region $|\eta| < 2.7$, but the innermost end-cap is replaced with CSCs to withstand the higher particle flux. CSCs are multiwire proportional chambers with cathode planes segmented into strips in orthogonal directions.

Resistive plate chambers (RPCs) and thin gap chambers (TGCs)
RPCs and TGCs provide less precise tracking, but at a faster readout speed needed for triggering. They also provide orthogonal coordinates to those of the MDTs and CSCs. RPCs are gas-filled parallel plate detectors operated in avalanche mode. TGCs are multiwire proportional chambers operated in quasi-saturated mode.

3.3.4 Trigger and Data Acquisition

It is technically infeasible to record the detector readout (1.6 MB per event) at the bunch crossing rate expected at the LHC (40 MHz). Moreover, the majority of these events are uninteresting in terms of the LHC physics program. For these reasons, ATLAS has a trigger system to identify and retain interesting events for further analysis offline.

3.3 The ATLAS Detector

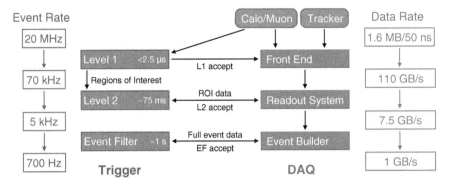

Fig. 3.8 Schematic diagram of the ATLAS trigger and data acquisition (DAQ) system. Typical values for event rates, data rates and trigger latencies during the 2012 run are taken from [4]

The trigger is split into three levels with successively lower event rates: level 1 (L1), level 2 (L2) and the event filter (EF). This is achieved by affording a longer latency for later levels to apply more refined selection criteria, as shown in Fig. 3.8. The L2 and EF triggers are collectively referred to as the high level trigger (HLT), and are implemented on an off-detector computer farm.

The L1 trigger searches for high-p_T leptons, photons and jets, in addition to events with significant total transverse energy or transverse energy imbalance. The decision is made using muon trigger chamber information (RPCs and TGCs) and reduced-granularity calorimeter information. Since the decision latency (2.5 μs) is longer than the bunch spacing (25 ns), event data is stored in pipelines while a decision is made.

The detector coordinates of interesting features identified by the L1 trigger, known as regions of interest (ROIs), are passed on to the L2 trigger. The L2 trigger then requests the full degree of detector information for these ROIs (∼2 % of total event data), in order to make a more informed decision. Finally, the EF trigger has access to the full event data, and further reduces the data rate such that it can be written to tape. The entire trigger system reduces the data rate from 64 TB/s to about 1 GB/s.

3.3.5 Detector Performance

The performance of each sub-detector is shown in Table 3.2. The σ_{p_T}/p_T of the tracker has a term proportional to p_T due to the intrinsic spatial resolution of the detector, and a constant term due to multiple scattering. The σ_E/E of the calorimeters has a constant term due to non-uniformities in the response, and a term proportional to $1/\sqrt{E}$ due to statistical fluctuations in the hadronic shower (the *stochastic term*).

Table 3.2 The measured performance in the central region of each of the ATLAS sub-detectors [5–8]

Detector component	Resolution	$|\eta|$ coverage	
		Measurement	L1 trigger
Tracker	$\sigma_{p_T}/p_T = 0.03\,\% \times p_T \oplus 1.5\,\%$	<2.5	–
EM calorimeter	$\sigma_E/E = 10\,\%/\sqrt{E} \oplus 1\,\%$	<3.2	<2.5
Hadronic calorimeter	$\sigma_E/E = 50\,\%/\sqrt{E} \oplus 3\,\%$	<4.9	<4.9
Muon spectrometer	$\sigma_{p_T}/p_T = 4\,\%$ at $p_T = 100$ GeV	<2.7	<2.4

Energies and momenta are given in GeV. The quoted MS performance is independent of the ID. The contribution of noise to σ_E/E (a term proportional to $1/E$) is neglected as it depends upon the pile-up environment

References

1. L. Evans, P. Bryant, LHC machine. JINST **3**, S08001 (2008). doi:10.1088/1748-0221/3/08/S08001
2. ATLAS Collaboration, Improved luminosity determination in pp collisions at $\sqrt{s}$ = 7 TeV using the ATLAS detector at the LHC, Eur. Phys. J. C **73**(1) (2013). doi:10.1140/epjc/s10052-013-2518-3, arXiv:1302.4393 [hep-ex]
3. ATLAS Collaboration, The ATLAS Experiment at the CERN Large Hadron Collider. JINST **3**, S08003 (2008). doi:10.1088/1748-0221/3/08/S08003
4. ATLAS Collaboration, K. Nagano, Algorithms, performance, and development of the ATLAS High-level Trigger, ATL-DAQ-SLIDE-2013-893 (2013) http://cds.cern.ch/record/1632445
5. ATLAS Collaboration, Alignment of the ATLAS inner detector and its performance in 2012, ATLAS-CONF-2014-047 (2014) http://cds.cern.ch/record/1741021
6. ATLAS Collaboration, Electron and photon energy calibration with the ATLAS detector using LHC Run 1 data (2014). arXiv:1407.5063 [hep-ex] (submitted to Eur. Phys. J. C)
7. ATLAS Collaboration, Jet momentum resolution with the ATLAS detector in proton-proton collisions at $\sqrt{s}$ = 8 TeV recorded in 2012, ATL-COM-PHYS-2014-010 (2014) (ATLAS internal) http://cds.cern.ch/record/1642375
8. ATLAS Collaboration, Measurement of the muon reconstruction performance of the ATLAS detector using 2011 and 2012 LHC proton-proton collision data (2014), arXiv:1407.3935 [hep-ex] (submitted to Eur. Phys. J. C)

Chapter 4
Overview of the $H \to WW$ Analysis

The WW decay of the Higgs boson is a promising search channel as it has a large branching ratio (BR) for a wide range of m_H. In fact, it is the most probable decay for $m_H > 135$ GeV (see Fig. 1.5). This chapter will describe the experimental search for $gg \to H \to WW \to \ell\nu\ell\nu$ (where $\ell = e, \mu$) using the 2012 dataset of pp collisions at $\sqrt{s} = 8$ TeV. The search strategy is optimised for a low mass Higgs boson, as favoured by electroweak fits, and thus accounts for off-shell W bosons.

The experimental signature of this search involves electrons, muons, jets and transverse momentum imbalance, and is outlined in Sect. 4.1. Then, Sect. 4.2 details how each of these objects is reconstructed by the ATLAS detector. Finally, Sect. 4.3 describes the criteria by which Higgs boson events are selected and background events are rejected. In doing so, it is necessary to make reference to many exclusive observables of signal and background processes. The modelling and estimation of these observables shall be described in Chaps. 5–7.

4.1 Experimental Signature

Following the $H \to WW$ decay, each W boson will decay leptonically or hadronically, with BR($W \to \ell\nu$) = 10.8 % and BR($W \to$ hadrons) = 67.6 % [1]. Thus, *dileptonic*, *semi-leptonic* and *hadronic* final states are conceivable. Although the dileptonic channel is suppressed by BRs, it is ultimately the most sensitive as the other two have larger backgrounds. This chapter will describe the dileptonic search, and henceforth 'lepton' and ℓ shall refer to an electron or muon.[1]

Electroweak fits favour a Higgs boson with mass $m_H < 2m_W$ (see Fig. 1.6). It is therefore important for the $H \to WW$ search to be sensitive to off-shell W

---

[1] Events with one or two $W \to \tau\nu$ decays can contribute to the dileptonic search when the τ decays to an electron or muon. This contribution is small however, since BR($\tau \to \ell\nu_\ell\nu_\tau$) = 17.6 % [1]. Also, the kinematics of such events are different due to the additional decay(s) and neutrinos.

Table 4.1 Summary of how each background produces the $\ell\ell + p_T^{inv}$ experimental signature, inclusive in the number of jets

Background	Mechanism of $\ell\ell + p_T^{inv}$ signature
WW	Irreducible
$t\bar{t}, tW$	Irreducible
tb, tbq	Jet fakes lepton
$Z/\gamma^* \to \ell\ell$	Fake p_T^{inv}
$Z/\gamma^* \to \tau\tau$	Irreducible (for leptonic τ decays)
$W +$ jet, dijet	Jet(s) fake lepton(s)
$W\gamma$	Photon fakes electron
$WZ, W\gamma^*, ZZ$	Unreconstructed lepton(s)

bosons. Experimentally, this means using leptons with low p_T thresholds, which unfortunately have reduced purity due to large misidentified hadronic backgrounds.

Since neutrinos do not interact with the detector, it is only possible to infer their combined transverse momentum from an imbalance in the visible momenta, called p_T^{inv}. Thus, it is not possible to fully reconstruct a mass peak in the $H \to WW \to \ell\nu\ell\nu$ search, so to be sensitive to the Higgs boson it becomes crucial to accurately understand the many background processes.

The basic experimental signature is two oppositely charged leptons and significant p_T^{inv}. However, there are background processes that exhibit the same signature. Others have an aspect of the signature faked by mismeasurement, or some part of their final state is not reconstructed. Table 4.1 introduces the different backgrounds. Jets are a convenient way to separate the contributions of different background processes.

Jets can also be used to separate the gluon-gluon fusion (ggF) and vector boson fusion (VBF) production modes of the Higgs boson (see Sect. 1.3). The search described below is designed for the ggF production mode, and this shall become particularly apparent when describing the $\geq$2-jet bin in Sect. 4.3.7.

4.2 Reconstruction of Physics Objects

Relating raw detector output to physical particles is an incredibly difficult task, particularly in a high pile-up environment. Sophisticated algorithms were developed for this task, and are outlined below.

4.2.1 Tracks

A track is a sequence of hits in the inner detector (ID) indicative of a charged particle trajectory. To first approximation the trajectories are helical, owing to the pervading solenoidal magnetic field. However, multiple scattering and energy losses can cause significant deviations from this path. Track reconstruction is possible in the region $|\eta| < 2.5$ (the extent of the ID).

4.2 Reconstruction of Physics Objects

The inputs to the track reconstruction are threefold: coordinates from the pixel and SCT,[2] and TRT drift circles. The *inside-out* algorithm [2, 3] uses a Kalman filter to seed tracks from hits in the three pixel layers and the first SCT layer. The seeds are then extended into the SCT and the TRT and fitted, whilst resolving ambiguities and applying quality criteria. Finally, the *outside-in* algorithm [2] considers unused track segments in the TRT, and extrapolates them into the SCT and pixel detector. This improves the tracking of secondary particles with a displaced vertex.

4.2.2 Primary and Secondary Vertices

A vertex is a location from which at least two outgoing tracks are reconstructed. *Primary vertices* are associated with the interactions of incoming protons, whereas *secondary vertices* are caused by particle decay or photon conversion (e^+e^- pair production).

Primary vertex reconstruction has two steps: association of tracks to vertices (*vertex finding*), and reconstruction of the vertex position itself (*vertex fitting*). ATLAS employs an iterative *finding-through-fitting* algorithm to simultaneously perform both steps [4, 5]. First, tracks originating from the interaction region ($\Delta z \approx \pm 5.6$ cm and $\Delta r \approx 15\,\mu$m) are identified and used to seed and fit a single vertex. Tracks considered outliers are then used to seed a new vertex, and a second fit of the two vertices is performed. The algorithm iterates, increasing the number of vertices, until the result stabilises.

The total number of primary vertices N_{PV} is used to assess the pile-up conditions of the bunch crossing. The primary vertex of the hard scatter is chosen to be that with the highest $\sum p_T^2$ of the constituent tracks, and is referred to as simply *the* primary vertex. As an additional quality criterion, this vertex must have at least three associated tracks.

Secondary vertex reconstruction is highly constrained by the physics of the vertex [5]. This can be enforced by mass or angular constraints, or in the track selection. For example, photon conversions are found using oppositely charged track pairs associated with electrons (via TRT identification), with small opening angle. Flavour tagging of jets, which uses secondary vertices, shall be described in Sect. 4.2.6.

4.2.3 Electrons

Electrons will pass through the ID before being absorbed by the electromagnetic calorimeter (ECal). They are therefore reconstructed by matching a track with an energy cluster in the ECal. Following reconstruction, the vast majority of elec-

---

[2]Each SCT module comprises two layers of silicon strips with a small stereo angle, enabling the three-dimensional coordinate to be precisely measured (see Sect. 3.3.1).

Fig. 4.1 The composition of reconstructed electron candidates as a function of transverse energy, simulated by a PYTHIA 6 sample of dijet, heavy flavour, prompt photon, W boson and Z boson production [6]. Non-isolated electrons are from heavy flavour decay. Background electrons are from photon conversions. Isolated electrons are prompt electrons from a W or Z boson, though appear below the y-axis range. The total integral is normalised to unity

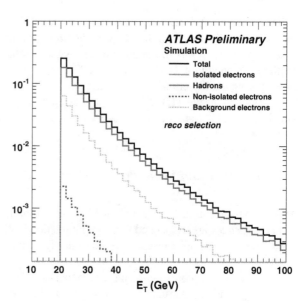

tron objects are actually misidentified hadrons, whilst many others are from photon conversions or electrons from heavy flavour decay (see Fig. 4.1). However, in the $H \rightarrow WW \rightarrow \ell\nu\ell\nu$ search, it is prompt electrons from the hard scatter that are of interest. Therefore identification and isolation criteria are applied to reject these backgrounds and select prompt electrons. The criteria were chosen to optimise the analysis sensitivity, e.g. a low E_T threshold of 10 GeV gives a large signal acceptance but must be compensated by tight identification requirements to reject the large W + jet and dijet backgrounds.

Electrons with $E_T > 10$ GeV and $|\eta| < 2.47$ are selected. Additionally, electrons with $1.37 < |\eta| < 1.52$ are vetoed since they correspond to the crack region (see Sect. 3.3.2).

Reconstruction

Energy deposits in ECal cells are clustered by a *sliding window* algorithm [6]. First, the calorimeter is divided into *towers* of size $\Delta\eta \times \Delta\phi = 0.025 \times 0.025$ (the cell size of the middle layer). Then a window of 3×5 towers in η-ϕ space scans the ECal for local maxima with $E_T > 2.5$ GeV.

Tracks within $\Delta R < 0.3$ of a cluster are then refit using a Gaussian sum filter (GSF) [7]. The default fit described in Sect. 4.2.1 uses a pion hypothesis to estimate material effects, and does not account for the significant bremsstrahlung experienced by electrons (which is highly η-dependent). The GSF is a non-linear fitter and improves the accuracy of electron tracking by accounting for this.

4.2 Reconstruction of Physics Objects

For a cluster to be reconstructed as an electron, it must be matched to a track. The matching uses an asymmetric $\Delta\phi$ requirement to allow for increased bending due to bremsstrahlung. If a track is within $\Delta\phi < 0.1(0.05)$ and $\Delta\eta < 0.05$ of the cluster centre, it is considered matched. When multiple tracks are matched, that with the smallest ΔR is chosen.

Finally, the cluster is rebuilt using a 3×7 (5×5) sliding window in the barrel (endcap), and the electron four-momentum is defined [8]. The direction is taken from the matched track and the energy is the cluster energy, corrected for losses in passive material and leakage outside the cluster. These simulation-based corrections depend on the longitudinal shower shape and energy deposited in the presampler. The EM energy scale was calibrated using test beam and in situ $Z \to ee$ measurements.

Identification

Electron identification is improved by applying cuts on track quality and shower shape variables. There are three reference operating points called *loose*, *medium* and *tight*, with progressively greater background rejection and lower signal efficiency.

For $E_T > 25$ GeV the medium criteria are used, comprising cuts on

- shower shape variables (lateral and longitudinal),
- leakage to the hadronic calorimeter (HCal),
- the number of pixel and SCT hits,
- the transverse impact parameter with respect to the primary vertex, d_0,
- track-cluster matching,
- transition radiation.

Additionally, electrons with conversion vertices or without a hit in the first pixel layer are vetoed (to suppress the $W\gamma$ and $W +$ jet backgrounds).

For $E_T < 25$ GeV the $W +$ jet and QCD backgrounds are much larger. Thus a multivariate *very tight likelihood* identification is used, with similar signal efficiency to the cut-based tight criteria but improved background rejection. It uses the signal and background probability density functions of multiple input variables to construct a likelihood discriminant, which may be cut upon (effectively choosing an operating point). The input variables are similar to those used in medium identification.

Isolation

Rejection of hadronic fakes or electrons from heavy flavour decays is further improved by requiring the electron to be isolated from activity in the tracker and calorimeter. The cuts are summarised in Table 4.2 and explained below.

Table 4.2 Tracker and calorimeter isolation criteria for electrons

Electron E_T (GeV)	Tracker isolation	Calorimeter isolation
10–15	$p_T^{\text{cone}}(0.4)/E_T < 0.06$	$E_T^{\text{cone}}(0.3)/E_T < 0.20$
15–20	$p_T^{\text{cone}}(0.3)/E_T < 0.08$	$E_T^{\text{cone}}(0.3)/E_T < 0.24$
>20	$p_T^{\text{cone}}(0.3)/E_T < 0.10$	$E_T^{\text{cone}}(0.3)/E_T < 0.28$

The cuts were chosen to optimise the sensitivity of a low mass $H \to WW \to \ell\nu\ell\nu$ search

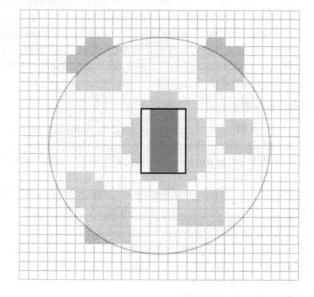

Fig. 4.2 Schematic of calorimeter isolation, adapted from reference [9]. *Pink cells* constitute topological clusters, the *yellow circle* is the isolation cone ($R_0 = 0.4$ shown), the 3×7 *green rectangle* is the reconstructed electron cluster, and the 5×7 *white rectangle* is the area removed from the cone

Tracker isolation: $p_T^{\text{cone}}(R_0)$ is the summed p_T of all tracks of $p_T > 0.4$ GeV within a cone of $\Delta R < R_0$, excluding the electron track itself. For robustness against pile-up, the tracks are required to originate from the primary vertex.

Calorimeter isolation: Uncalibrated topological clusters (see Sect. 4.2.5) are built from energy deposits in the ECal and HCal. These are more robust against pile-up than cell deposits as they suppress noise. $E_T^{\text{cone}}(R_0)$ is the summed E_T of the topological clusters within a cone of $\Delta R < R_0$, excluding cells in a 5×7 window around the electron (see Fig. 4.2). Corrections are made for electron energy leakage and deposits from pile-up and the underlying event.

Quality

An electron is rejected if its cluster is affected by a localised detector problem (*e.g.* a dead cell).

Primary vertex association

To associate the electron with the primary vertex, the transverse impact parameter d_0 is required to be within three standard deviations of zero. Also, the longitudinal impact parameter z_0 is constrained by $|z_0 \sin\theta| < 0.4$ mm.

4.2 Reconstruction of Physics Objects

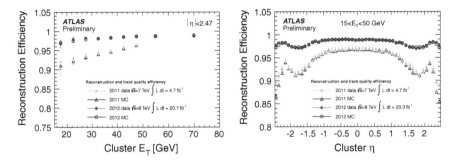

Fig. 4.3 Electron reconstruction efficiency versus E_T (*left*) and η (*right*), measured using tag-and-probe of $Z \to ee$ data and compared to MC [10]. In 2012, the implementation of GSF tracking gave significant performance improvement

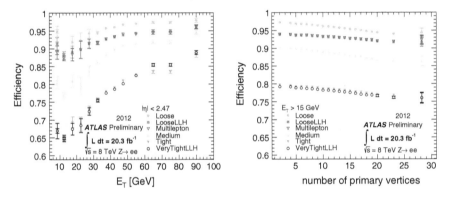

Fig. 4.4 Efficiency of a variety of electron identification operating points versus E_T (*left*) and N_{PV} (*right*), measured using tag-and-probe of $Z \to ee$ data [10]. 'Very tight likelihood' has similar efficiency to 'tight', but improves background rejection. All operating points are fairly robust against pile-up. 'Multilepton' is an operating point developed for multilepton searches, and uses observables related to bremsstrahlung

Efficiency

The efficiency of each selection step (reconstruction, identification, isolation and primary vertex association) is measured via *tag-and-probe* of $Z \to ee$ events [8, 10]. This involves selecting an unbiased sample of events containing a well-identified *tag* object and a loosely-identified *probe* object, and measuring the selection efficiency of the probes. The sample must be clean (often enforced by a mass constraint) and backgrounds estimated (usually with a side-bands or template fit). The tag is a well-identified electron, the probe is an electron passing the previous step (or an ECal cluster) and the constraint is $80\,\text{GeV} < m_{(\text{tag, probe})} < 100\,\text{GeV}$. The reconstruction and identification efficiencies are shown in Figs. 4.3 and 4.4 respectively. Comparison with MC yields efficiency scale factors.

Table 4.3 Tracker and calorimeter isolation criteria for muons

Muon p_T (GeV)	Tracker isolation	Calorimeter isolation
10–15	$p_T^{cone}(0.4)/p_T < 0.06$	$E_T^{cone}(0.3)/p_T < 0.06$
15–20	$p_T^{cone}(0.3)/p_T < 0.08$	$E_T^{cone}(0.3)/p_T < 0.12$
20–25	$p_T^{cone}(0.3)/p_T < 0.12$	$E_T^{cone}(0.3)/p_T < 0.18$
>25	$p_T^{cone}(0.3)/p_T < 0.12$	$E_T^{cone}(0.3)/p_T < 0.30$

The cuts were chosen to optimise the sensitivity of a low mass $H \to WW \to \ell\nu\ell\nu$ search

4.2.4 Muons

Muons will pass through the ID, deposit minimal energy in the calorimeters and continue through the muon spectrometer (MS). Thus, they can be found via activity in the MS, optionally matched to an ID track. Muons with $p_T > 10$ GeV and $|\eta| < 2.5$ are selected.

Reconstruction

Muons are found by matching MS tracks to ID tracks[3] and their momenta are measured from a combination of the track curvatures [3].

First, the *Muonboy* algorithm [11] identifies a $\Delta\eta \times \Delta\phi \approx 0.4 \times 0.4$ region of activity via the muon trigger chambers (see Sect. 3.3.3), and then reconstructs localised track segments from nearby MDT and CSC hits. Segments in different layers are combined to form a track. Finally, a global track fit using the full MS hit information is performed, accounting for energy loss.

MS tracks are extrapolated to the primary vertex, accounting for energy loss in the calorimeter, and matched to ID tracks (see Sect. 4.2.1). The statistical combination of the two tracks is performed by the *STACO* algorithm [11], which weights each track by its covariance matrix. The muon four-momentum is determined from this combined track (dominated by the ID track at low-p_T and by the MS track at high-p_T), and is calibrated with $Z \to \mu\mu$ events [12].

Isolation

Although the muon signature is much cleaner than that of electrons, isolation criteria are still required to suppress muons from hadronic decays and misidentified hadrons exiting the calorimeter (*punch-through* or *sail-through*). These criteria are summarised in Table 4.3. The $p_T^{cone}(R_0)$ and $E_T^{cone}(R_0)$ variables are defined similarly to electrons, though the size of the subtracted calorimeter window is reduced for muons.

---

[3] A variety of muon reconstruction strategies are available: *stand-alone* muons have only MS tracks, *combined* muons have both MS and ID tracks, *segment-tagged* muons have ID tracks that match to an MS track segment, and *calorimeter-tagged* muons have ID tracks matched to the calorimeter deposit of a minimum ionising particle (used to recover efficiency at $\eta \approx 0$, where detector services reduce the MS coverage). This thesis uses *combined* muons, which offer the best performance.

4.2 Reconstruction of Physics Objects

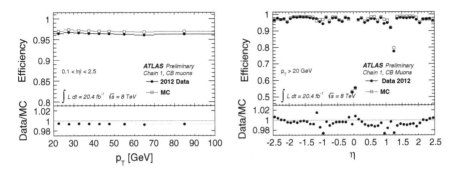

Fig. 4.5 Muon reconstruction efficiency versus p_T (*left*) and η (*right*), measured using tag-and-probe of $Z \to \mu\mu$ data and compared to MC [12]. Reductions in efficiency at $\eta \approx 0$ and $\eta \approx 1.2$ are due to partial MS coverage (detector services and incomplete installation respectively). 'Chain 1' refers to *STACO* reconstruction

Quality

The ID track must satisfy quality criteria on the number of traversed sensors:

- $n_{\text{pixel}}^{\text{hit}} + n_{\text{pixel}}^{\text{dead}} \geq 1$
- $n_{\text{SCT}}^{\text{hit}} + n_{\text{SCT}}^{\text{dead}} \geq 5$
- $n_{\text{pixel}}^{\text{hole}} + n_{\text{SCT}}^{\text{hole}} \leq 2$
- $n_{\text{TRT}}^{\text{hit}} + n_{\text{TRT}}^{\text{outlier}} \geq 6$ and, for $|\eta| < 1.9$, at least 10% are hits

where a hole is an expected hit that is not observed (excluding dead modules).

Primary vertex association

To associate the muon with the primary vertex, the transverse impact parameter d_0 is required to be within three standard deviations of zero. Also, the longitudinal impact parameter z_0 is constrained by $|z_0 \sin\theta| < 1.0\,\text{mm}$.

Efficiency

The efficiency of each selection step is measured via tag-and-probe (see Sect. 4.2.3) of $Z \to \mu\mu$ events [12]. For example, the reconstruction efficiency is shown in Fig. 4.5. Comparison with MC yields efficiency scale factors.

4.2.5 Jets

A hadron will pass through the ID (leaving hits if charged) before being absorbed by the ECal and HCal. The finite resolution of the calorimeter and the large number of overlapping hadronic showers make it impossible to reconstruct individual hadrons. For this reason, and more theoretical ones outlined in Sect. 2.3, jets are reconstructed from energy deposits. The main challenge in identifying jets originating from the hard scatter is calorimeter noise resulting from the electronics and pile-up.

Jets are selected with $p_T > 25$ GeV for central jets ($|\eta| < 2.4$) and with $p_T > 30$ GeV for forward jets ($2.4 < |\eta| < 4.5$). However, occasionally it is useful to use jets with other p_T thresholds (*e.g.* the central jet veto in the VBF search). It shall be clearly stated when this is the case.

Reconstruction

To suppress contributions from calorimeter noise, energy deposits are grouped into topological clusters (*topo-clusters*) [13, 14]. First, a seed cell is identified with signal-to-noise ratio $S/N > 4$. Surrounding cells with $S/N > 2$ are iteratively added. Finally, a single iteration of neighbouring cells with no S/N criterion are added. When localised maxima exist within a cluster, it may be split. Clearly, the robustness to noise is determined by the choice of N, which is taken to be the sum in quadrature of the electronic noise and the pile-up noise expected with $\mu = 30$. The non-compensating calorimeter design yields different responses to electromagnetic and hadronic showers.[4] Topo-clusters are reconstructed with an assumed response of ≈ 1 (EM scale) and should be corrected for average energy loss in nuclear interactions (jet energy scale or JES). An intermediate *local cluster weighting* (LCW) classifies each topo-cluster as electromagnetic or hadronic (via energy density and shower depth) and applies an appropriate energy correction.

These topo-clusters are then input to the anti-k_T jet algorithm with $R = 0.4$ (see Sect. 2.3), under the assumption that the topo-clusters are massless.

Calibration

There are four steps to the jet calibration [14], described below.

First, the jet direction is adjusted such that it originates from the primary vertex, rather than the centre of the ATLAS coordinate system (a remnant of the jet algorithm).

Second, energy contributed by pile-up is subtracted [15]. This involves measuring the jet area A (its susceptibility to additional particles with infinitesimal p_T) and the pile-up density ρ of the event (the median p_T/A of all k_T jets with $R = 0.4$). Residual $|\eta|$-dependent corrections for in-time pile-up, $\alpha(N_{PV} - 1)$, and out-of-time pile-up, $\beta(\mu)$, are also made. Thus, the correction is

$$p_T^{\text{corr}} = p_T - \rho A - \alpha(N_{PV} - 1) - \beta(\mu) \tag{4.1}$$

and the improved robustness to pile-up that it provides is shown in Fig. 4.6.

Third, the energy is calibrated from the LCW scale to the JES. The correction is derived from MC by comparing the energy of a reconstructed jet to that of the corresponding hadron-level jet (*i.e.* before detector simulation).

---

[4]Significant hadronic shower energy is lost through slow neutrons, nuclear excitations and neutrinos.

4.2 Reconstruction of Physics Objects

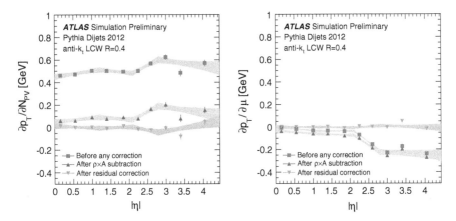

Fig. 4.6 Dependence of jet p_T on in-time pile-up (*left*) and out-of-time pile-up (*right*), versus $|\eta|$ [15]. The bands show the fits used in the correction

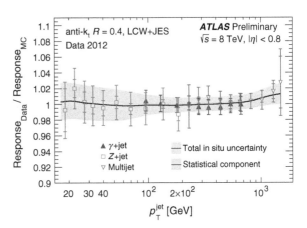

Fig. 4.7 The ratio of the jet response measured in data to the MC prediction versus jet p_T [16]. This is the inverse of the in situ JES correction (see text). The results of three p_T balance studies and their combination (*black line*) are shown

Finally, a residual in situ JES calibration corrects for mismodelling, and is applied to jets in data events only. The calibration exploits the p_T balance between a jet and a well-measured reference object, deriving the correction as the double-ratio

$$\left\langle p_T^{\text{jet}}/p_T^{\text{ref}} \right\rangle_{\text{MC}} \Big/ \left\langle p_T^{\text{jet}}/p_T^{\text{ref}} \right\rangle_{\text{data}} . \tag{4.2}$$

The reference object is chosen to be a Z boson (decaying to e^+e^- or $\mu^+\mu^-$) at low p_T, a photon at medium p_T, or a system of well-calibrated low-p_T jets at high p_T. The contribution of each technique to the correction is shown in Fig. 4.7.

Quality

Quality criteria (known as *looser*) reject calorimeter noise based on pulse shapes [17]. The selection is more than 99.8 % efficient for jets with $p_T > 20$ GeV.

Primary vertex association

The *jet vertex fraction* (JVF) is defined as the fraction of associated tracks originating from the primary vertex ($|z_0 \sin\theta| < 1$ mm), weighted by track p_T [15]. It is a powerful discriminant against pile-up jets, but is only available for $|\eta| < 2.4$. Central jets with $p_T < 50$ GeV are required to have either JVF > 0.5 or zero associated tracks. The efficiency of this cut is measured in situ with $Z +$ jet events.

4.2.6 b-jets

A *b-tagging algorithm* exploits the long lifetimes of b-hadrons to identify jets originating from b-quarks. This aids suppression of top backgrounds, since BR($t \to Wb$) ≈ 100 %. The MV1 algorithm uses a neural network with inputs of jet p_T and η, and the weights output by three other algorithms [18]:

- IP3D identifies jet tracks with significant impact parameters,
- SV1 searches for secondary vertices associated with b-hadron decays,
- JetFitter reconstructs the decay chain topology (including the daughter c-hadron).

An operating point is chosen to tag 85 % of b-jets, with a mis-tag rate of $\sim$40 % for c-jets, $\sim$35 % for τ-jets and $\sim$10 % for light jets (initiated by u, d, s or g). Note that b-tagging is only available for jets with $|\eta| < 2.5$, since it relies upon tracks.

Modelling b-tagging algorithms is difficult, partly due to limited knowledge of b-hadron decays; thus it is imperative to measure their efficiency with experimental data. This was done with dileptonic $t\bar{t}$ decays, using a combinatorial likelihood method [19]. Comparison with MC yields efficiency scale factors.

4.2.7 Missing Transverse Momentum

The near-hermetic detector design enables information of non-interacting particles (*e.g.* neutrinos) to be inferred from measurements of interacting particles. Since the initial state has negligible nett $\boldsymbol{p}_T$, momentum conservation implies that the visible and invisible $\boldsymbol{p}_T$ in the final state must balance. Thus, the negative sum of the visible $\boldsymbol{p}_T$ is known as *missing transverse momentum*, and equals the sum of the invisible $\boldsymbol{p}_T$. Its magnitude is referred to as p_T^{inv}, though alternative symbols are used depending upon its experimental reconstruction (see below).

Since the measurement of p_T^{inv} relies on the measurement of all the visible particles produced in the event, it generally has a poor experimental resolution. In particular, pile-up has an adverse effect on performance (see Fig. 4.8). Object mismeasurement can produce a fake imbalance that is highly correlated to the event kinematics.

4.2 Reconstruction of Physics Objects

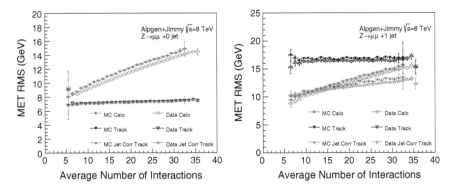

Fig. 4.8 The root-mean-square of the missing transverse momentum versus the mean number of interactions per bunch crossing, μ, measured in $Z \to \mu\mu$ events featuring 0 jets (*left*) and 1 jet (*right*). Results are shown for calorimeter-based E_T^{miss} (*red*), track-based p_T^{miss} (*black*) and jet-corrected track-based $p_T^{miss,corr}$ (*blue*). In the 0-jet case, p_T^{miss} and $p_T^{miss,corr}$ are identical

Calorimeter-based E_T^{miss}

One can reconstruct missing transverse momentum with calorimeter energy deposits [20]. This is denoted $\boldsymbol{E}_T^{miss}$, with magnitude E_T^{miss}, since it is calorimeter-based.

First, calorimeter cells are associated with, and replaced by, calibrated high-p_T objects. To avoid double counting, the association is done in a specific order: electrons, photons [21], hadronically decaying τ-leptons [22], jets and muons. For this measurement, some object selection criteria are loosened with respect to the preceding descriptions. Importantly, the JVF criterion is removed from jets and their p_T threshold is lowered to 20 GeV. Also, segment-tagged muons are used in addition to combined muons.

The low-p_T component is reconstructed using topo-clusters of the remaining calorimeter deposits, which are calibrated to the LCW scale (see Sect. 4.2.5). This is further improved by including unassociated tracks with $p_T > 400$ MeV. Thus, the calorimeter-based version of p_T^{inv} is defined as

$$E_T^{miss} = -\left\{ \sum_e \boldsymbol{p}_T + \sum_\mu \boldsymbol{p}_T + \sum_\tau \boldsymbol{p}_T + \sum_\gamma \boldsymbol{p}_T + \sum_{jets} \boldsymbol{p}_T + \sum_{soft} \boldsymbol{p}_T \right\}. \quad (4.3)$$

Unfortunately, E_T^{miss} is highly susceptible to pile-up because the starting points are calorimeter cells, which are not easily associated to the primary vertex. This is shown by the red points in Fig. 4.8.

Track-based p_T^{miss}

One can alternatively reconstruct missing transverse momentum with ID tracks. This is denoted $\boldsymbol{p}_T^{miss}$, with magnitude p_T^{miss}, since it is track-based.

Tracks with $p_T > 500\,\text{MeV}$ are selected if they have sufficiently small impact parameters with respect to the primary vertex ($|d_0| < 1.5\,\text{mm}$ and $|z_0 \sin\theta| < 1.5\,\text{mm}$). Quality criteria require $n_{\text{pixel}}^{\text{hit}} \geq 1$ and $n_{\text{SCT}}^{\text{hit}} \geq 6$. Additional tracks may be used if they are associated with a lepton with $|z_0 \sin\theta| < 1.0\,\text{mm}$ (the electron identification criteria are loosened to medium c.f. Sect. 4.2.3, and the muon p_T threshold is lowered to 6 GeV c.f. Sect. 4.2.4).

Occasionally, tracks are poorly reconstructed, which can have a significant impact upon p_T^{miss}. Therefore, tracks with $p_T > 100\,\text{GeV}$ that are not associated to any physics object are removed from the selection. Similarly, a track within $\Delta R < 0.4$ of a jet with $p_T > 10\,\text{GeV}$ is removed if $p_T^{\text{track}} > 1.4\,p_T^{\text{jet}}$.

Bremsstrahlung from an electron can convert to e^+e^- pairs, resulting in additional tracks. For this reason, tracks within a cone of $\Delta R < 0.05$ of a reconstructed electron are collectively replaced with the calibrated electron p_T. Muon ID tracks are also replaced by the p_T of the reconstructed muon.
Thus, the track-based version of p_T^{inv} is defined as

$$p_T^{\text{miss}} = -\left\{ \sum_e \boldsymbol{p}_T + \sum_\mu \boldsymbol{p}_T + \sum_{\text{unassoc.}} \boldsymbol{p}_T \right\}. \tag{4.4}$$

p_T^{miss} is very resilient to pile-up. However, its inability to include neutral hadrons significantly degrades the resolution. This is shown by the black points in Fig. 4.8.

Jet-corrected track-based $p_T^{\text{miss,corr}}$

It is possible to correct p_T^{miss} for neutral hadrons using calorimeter information. Tracks within $\Delta R < 0.4$ of a jet are collectively replaced by the calibrated jet p_T. Thus, the jet-corrected track-based version of p_T^{inv} is defined as

$$p_T^{\text{miss,corr}} = -\left\{ \sum_e \boldsymbol{p}_T + \sum_\mu \boldsymbol{p}_T + \sum_{\text{jets}} \boldsymbol{p}_T + \sum_{\text{unassoc.}} \boldsymbol{p}_T \right\}. \tag{4.5}$$

This has the best resolution, but the use of calorimeter information does introduce some pile-up dependence. This is shown by the blue points in Fig. 4.8.

Relative missing transverse momentum

The effect of object mismeasurement upon missing transverse momentum can be reduced by considering the component transverse to the nearest reconstructed object. Thus, the relative E_T^{miss} is defined to be

$$E_{T,\text{rel}}^{\text{miss}} = \begin{cases} E_T^{\text{miss}} \sin(\Delta\phi) & \text{if } \Delta\phi < \pi/2 \\ E_T^{\text{miss}} & \text{otherwise} \end{cases} \tag{4.6}$$

where $\Delta\phi$ is the azimuthal angle between E_T^{miss} and the nearest electron, muon or jet object. Similar definitions exist for $p_{T,\text{rel}}^{\text{miss}}$ and $p_{T,\text{rel}}^{\text{miss,corr}}$.

4.2.8 Object Overlap Removal

In order to avoid double-counting calorimeter deposits and tracks as multiple physics objects, overlapping objects are removed according to the following rules:
- $\Delta R(e, \mu) < 0.1$ remove electron
- $\Delta R(e, e) < 0.1$ remove electron with lower p_T
- $\Delta R(e, j) < 0.3$ remove jet
- $\Delta R(\mu, j) < 0.3$ remove muon

where e, μ and j are the electron, muon and jet objects defined in Sects. 4.2.3–4.2.5.

Additionally, events are rejected if a muon passing loose reconstruction (*i.e.* without isolation, quality, primary vertex and p_T criteria) is found within $\Delta R < 0.05$ of an electron passing the full reconstruction. This is effective at rejecting rare $Z \to \mu\mu$ events where a muon radiates a hard photon, which subsequently converts and passes the electron reconstruction.

4.3 Event Selection Criteria

Section 4.1 outlined the basic experimental signature of the search as two oppositely charged leptons and significant p_T^{inv}. Thus, the initial stages of the event selection are responsible for finding this signature. Subsequent criteria, or *cuts*, target specific background processes. Their aim is to improve the analysis sensitivity by suppressing backgrounds whilst retaining a sufficient number of signal events.

4.3.1 Data Quality

The *pp* dataset (see Sect. 3.2.2) is hierarchically split into *periods* of broadly consistent beam conditions, *runs* typically corresponding to LHC fills, and *luminosity blocks* of ∼2 min where the instantaneous luminosity is approximately constant. Luminosity blocks are included in the analysis if the detector was operating sufficiently for the recorded data to be considered 'good for physics' (see Fig. 3.2a).[5] For the 2012 dataset, this corresponds to a total integrated luminosity of $20.3\,\text{fb}^{-1}$.

Individual events are also vetoed if certain data quality criteria are failed by:

- a noise burst in the LAr calorimeter,
- data corruption caused by a restart of the synchronisation system,
- a *looser* jet is reconstructed with $p_T > 20\,\text{GeV}$ (indicative of an HCal spike),
- a jet is reconstructed near a 'hot' HCal tile (1st–8th May 2012 only).

---

[5] ATLAS good runs list: `data12_8TeV.periodAllYear_DetStatus-v61-pro14-02_DQDefects-00-01-00_PHYS_StandardGRL_All_Good.xml`.

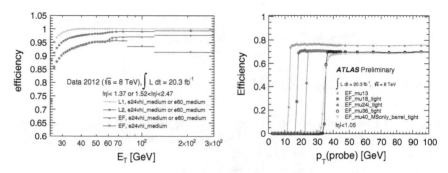

Fig. 4.9 Efficiencies of the single-lepton triggers for electrons with respect to offline *medium* identification (*left*) and muons with respect to offline reconstruction (*right*) [23]

A further quality criterion requires that the primary vertex considered as the hard scatter (that with the highest $\sum p_T^2$) must be associated with at least three tracks. This reduces the cosmic ray background to negligible levels.

4.3.2 Trigger

It is infeasible to record all the delivered collisions; ATLAS employs a trigger system to identify and record interesting events (see Sect. 3.3.4). In the $H \to WW \to \ell\nu\ell\nu$ search it is natural to trigger on high-p_T leptons, using algorithms similar to, though less sophisticated than, those in Sects. 4.2.3 and 4.2.4.

A trigger is characterised by its efficiency versus p_T curve (though it also depends on η), which has a turn-on region followed by a plateau, as shown in Fig. 4.9. It is preferable to operate on the plateau, where the efficiency is more stable and has smaller uncertainty. To maximise the signal yield, it is desirable to use a trigger with a lower turn-on p_T. However, increased backgrounds and limitations to trigger latency and bandwidth require a compromise to be found. The lowest unprescaled[6] single-lepton triggers available in 2012 had nominal p_T thresholds of 24 GeV. Fortunately, it is possible to recover trigger efficiency at lower p_T by using dilepton triggers, because the backgrounds are much smaller when two leptons are required.

Events are required to pass at least one trigger listed in Table 4.4. The single-lepton triggers include a tighter low-p_T trigger and a looser high-p_T trigger in order to maximise the efficiency. Dilepton triggers are then used to recover some efficiency at lower p_T. Together, these triggers support a dilepton signature with p_T thresholds of 22 and 10 GeV in the offline analysis, whilst operating on the plateau.

Additionally, events are required to have at least one lepton passing the offline reconstruction that is matched within $\Delta R < 0.15$ of a triggered lepton object. Single-

---

[6]A *prescaled* trigger reduces the threshold p_T by recording only 1 in N events passing the trigger, and weighting such events by a factor N. In doing so, statistical power is lost.

4.3 Event Selection Criteria

Table 4.4 Employed triggers

Single-lepton triggers	e	EF_e24vhi_medium1 or EF_e60_medium1
	μ	EF_mu24i_tight or EF_mu36_tight
Dilepton triggers	ee	EF_2e12Tvh_loose1 or EF_2e12Tvh_loose1_L2StarB
	μμ	EF_mu18_tight_mu8_EFFS
	eμ	EF_e12Tvh_medium1_mu8

EF refers to event filter, e is an electron, mu is a muon, the subsequent number is the p_T threshold, vh indicates calorimeter isolation, i indicates track isolation, and tight, medium or loose is the identification. Other parts relate to the trigger chain. Criteria are looser than those applied offline

lepton triggers are matched to offline leptons with $p_T > 25$ GeV. Dilepton triggers comprise two of the following single-lepton triggers: mu8 is matched to offline muons with $p_T > 10$ GeV, mu18 is matched to offline muons with $p_T > 20$ GeV, and e12 is matched to offline electrons with $p_T > 15$ GeV.[7]

Lepton trigger efficiencies are measured via tag-and-probe (see Sect. 4.2.3) of $Z \to \ell\ell$ events, where the tag and probe have both passed the offline lepton selection and the tag has successfully matched to a triggered lepton object. For example, single-lepton trigger efficiencies are displayed in Fig. 4.9. Comparison with MC yields efficiency scale factors.

4.3.3 Pre-selection of Dilepton + p_T^{inv} Signature

Following the trigger selection, events are required to have two oppositely charged leptons passing the offline selection (see Sect. 4.2). The lepton with the highest p_T, called the *leading lepton*, must have $p_{T,\ell}^{\text{lead}} > 22$ GeV and the *subleading lepton* must have $p_{T,\ell}^{\text{sublead}} > 10$ GeV. Events containing a third lepton are vetoed in order to reject backgrounds with three or more leptons in the final state, such as WZ production.

At this point, it is possible to split the dilepton final state into four channels according to the flavours of the two leptons: ee, $\mu\mu$, $e\mu$, μe (where the first flavour is that of the leading lepton). This is very useful because the background compositions of the channels are dramatically different. For example, the $Z/\gamma^* \to \ell\ell$ background is much larger in the same flavour channels ($ee/\mu\mu$) than the different flavour channels ($e\mu/\mu e$).

Low mass hadronic resonances with dileptonic decays (*e.g.* J/ψ and Υ) are removed from the $ee/\mu\mu$ channels by requiring the mass of the dilepton system $m_{\ell\ell} > 12$ GeV. This also greatly suppresses the low mass Z/γ^* background. For the $e\mu/\mu e$ channels, a cut of $m_{\ell\ell} > 10$ GeV suppresses leptons from heavy flavour

---

[7] It follows that events featuring an electron with $p_T < 15$ GeV must fire a single-lepton trigger, and thus the leading lepton must have $p_T > 25$ GeV.

58   4 Overview of the $H \rightarrow WW$ Analysis

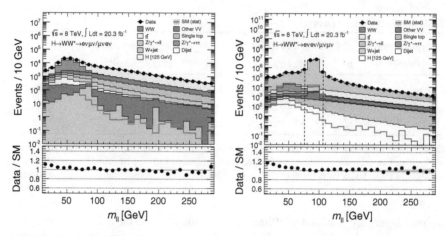

Fig. 4.10 The $m_{\ell\ell}$ distribution in the $e\mu/\mu e$ (*left*) and $ee/\mu\mu$ (*right*) channels. These are made after the low $m_{\ell\ell}$ cut. The *dashed lines* enclose the $|m_{\ell\ell} - m_z| > 15$ GeV cut applied to the $ee/\mu\mu$ channels

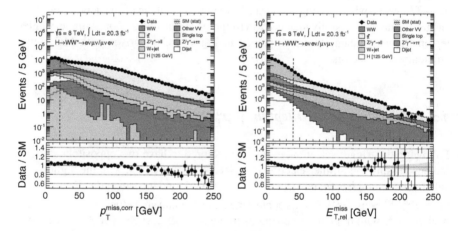

Fig. 4.11 The p_T^{inv} observable used in the selection of the $e\mu/\mu e$ (*left*) and $ee/\mu\mu$ (*right*) channels. These are made after the Z mass veto cut. The selection removes events to the left of each *dashed line*

decays and resonance decays to τ leptons. The $ee/\mu\mu$ channels are dominated by the Z/γ^* background (see Fig. 4.10), but most of these events can be rejected by vetoing a window around the Z mass, $|m_{\ell\ell} - m_z| > 15$ GeV, where $m_z = 91.1876$ GeV.

Requiring significant p_T^{inv} suppresses the $Z/\gamma^* \rightarrow \ell\ell$ and dijet backgrounds (see Fig. 4.11). It also suppresses the $Z/\gamma^* \rightarrow \tau\tau$ background as its neutrinos have a propensity to cancel in the p_T^{inv} calculation. We require $p_T^{\mathrm{miss,corr}} > 20$ GeV in the $e\mu/\mu e$ channels and $E_{T,\mathrm{rel}}^{\mathrm{miss}} > 40$ GeV in the $ee/\mu\mu$ channels. In the $e\mu/\mu e$ channels, the p_T^{inv} cut is relaxed because the $Z/\gamma^* \rightarrow \ell\ell$ background is smaller.

4.3 Event Selection Criteria

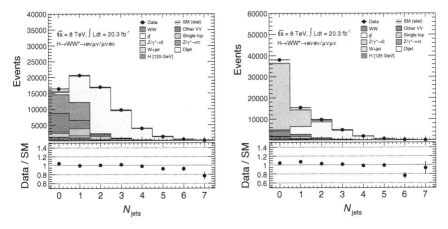

Fig. 4.12 Jet multiplicity distribution in the $e\mu/\mu e$ (*left*) and $ee/\mu\mu$ (*right*) channels. These are made following the pre-selection

Following the selection of dilepton events with significant p_T^{inv}, the $e\mu/\mu e$ channels are dominated by top background and the $ee/\mu\mu$ channels are dominated by $Z/\gamma^* \to \ell\ell$. However, the background compositions of all channels are highly dependent upon the number of jets (see Fig. 4.12). Thus, at this stage the analysis is binned according to jet multiplicity (0-jet, 1-jet, $\geq$2-jet), so that backgrounds can be targeted individually. In the $\geq$2-jet bin, only the $e\mu/\mu e$ channels are used.

4.3.4 $H \to WW \to \ell\nu\ell\nu$ Decay Topology

The discrimination of $H \to WW$ signal from irreducible backgrounds that also feature a WW pair is a problem common to all jet bins. Thus, the topology of the $H \to WW \to \ell\nu\ell\nu$ decay is discussed before continuing with the event selection.

First, the spin-0 nature of the Higgs boson and the V−A structure of the weak interaction (see Sect. 1.2) imply that a small opening angle between the two leptons is preferred. This follows from spin conservation and the extremely small masses of the neutrinos. Consequently, the mass of the dilepton system, $m_{\ell\ell}$, is also small. This follows from $m_{\ell\ell}^2 \simeq 2E_{\ell_1}E_{\ell_2}(1-\cos\vartheta)$, where ϑ is the opening angle. Thus, signal events are selected with the criteria $\Delta\phi(\ell,\ell) < 1.8$ and $m_{\ell\ell} < 55\,\text{GeV}$.

Second, the invariant mass of the dilepton + dineutrino system should correspond to m_H (modified by a Breit-Wigner distribution). Unfortunately, as discussed above, it is only possible to infer the *transverse* component of the dineutrino system momentum. We therefore construct a transverse mass variable

$$m_T = \sqrt{(E_{T,\ell\ell} + p_T^{\text{miss,corr}})^2 - \left|\boldsymbol{p}_{T,\ell\ell} + \boldsymbol{p}_T^{\text{miss,corr}}\right|^2} \qquad (4.7)$$

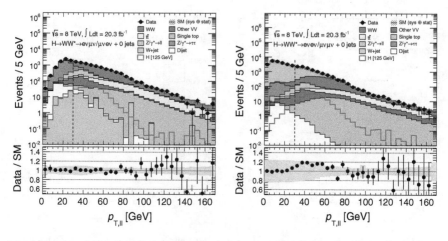

Fig. 4.13 The $p_{T,\ell\ell}$ distribution in the $e\mu/\mu e$ (*left*) and $ee/\mu\mu$ (*right*) channels. These are made in the 0-jet bin, after the $\Delta\phi(\ell\ell, p_T^{inv})$ cut. The selection removes events to the left of each *dashed line*

where $E_{T,\ell\ell}^2 = p_{T,\ell\ell}^2 + m_{\ell\ell}^2$ and $\boldsymbol{p}_T^{miss,corr}$ is used as it has the best resolution (see Sect. 4.2.7). At hadron-level this has an upper bound at m_H, though at detector-level the sharp cut-off is smeared by the poor $p_T^{miss,corr}$ resolution. This m_T observable is used in the statistical fitting procedure.

4.3.5 0-jet Selection

The 0-jet bin is dominated by Z/γ^* and WW backgrounds. A small number of pathological events where $\boldsymbol{p}_T^{miss,corr}$ is near $\boldsymbol{p}_{T,\ell\ell}$ are rejected by requiring $\Delta\phi(\ell\ell, p_T^{inv}) > \pi/2$.

Considering the $Z/\gamma^* \to \ell\ell$ background, the boson p_T will generally be less than the p_T threshold used in jet selection, since no jet has been found. Thus, the Z/γ^* background is greatly reduced by requiring $p_{T,\ell\ell} > 30\,\text{GeV}$ (see Fig. 4.13). In the $ee/\mu\mu$ channels only, the $Z/\gamma^* \to \ell\ell$ background is also suppressed by an additional cut $p_{T,rel}^{miss} > 40\,\text{GeV}$ (see Fig. 4.14).

After the signal topology selection (see Sect. 4.3.4), a large $Z/\gamma^* \to \ell\ell$ background remains in the $ee/\mu\mu$ channels, despite the significant p_T^{inv} requirement. This can be further suppressed by searching for soft hadronic activity to balance the dilepton system. First, jets with $p_T > 10\,\text{GeV}$ and $|\eta| < 4.5$ are found (as detailed in Sect. 4.2.5, minus the JVF cut). Then a discriminant is defined as

$$f_{\text{recoil}} = \left| \sum_{j \text{ in } \wedge} \text{JVF}_j \cdot \boldsymbol{p}_{T,j} \right| \Big/ p_{T,\ell\ell} \quad (4.8)$$

4.3 Event Selection Criteria

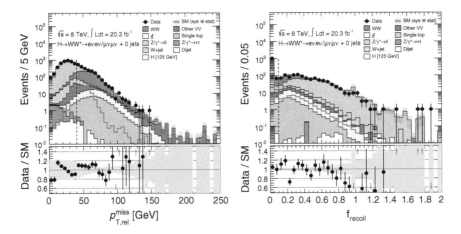

Fig. 4.14 The $p_{\mathrm{T,rel}}^{\mathrm{miss}}$ (*left*) and f_{recoil} (*right*) distributions in the $ee/\mu\mu$ channels. The subsequent selection removes events to the left of the *dashed line* in the $p_{\mathrm{T,rel}}^{\mathrm{miss}}$ plot, and removes events to the right of the *dashed line* in the f_{recoil} plot

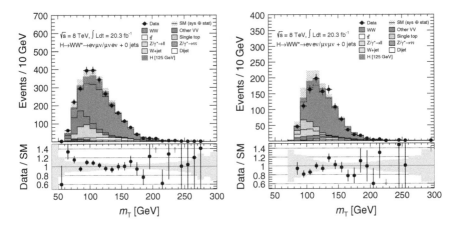

Fig. 4.15 The m_{T} distribution of the selected 0-jet events, in the $e\mu/\mu e$ (*left*) and $ee/\mu\mu$ (*right*) channels

where $\wedge$ is the detector quadrant centred on $-\boldsymbol{p}_{\mathrm{T},\ell\ell}$. This is essentially the fraction of $p_{\mathrm{T},\ell\ell}$ that can be balanced by soft hadronic activity in the opposing quadrant, and so is larger in $Z/\gamma^* \to \ell\ell$ than in processes featuring prompt neutrinos. We require $f_{\mathrm{recoil}} < 0.1$ in the $ee/\mu\mu$ channels (see Fig. 4.14). f_{recoil} is also instrumental in estimating the $Z/\gamma^* \to \ell\ell$ background, and shall be revisited in Sect. 7.4.

The m_{T} distributions of the selected 0-jet events are shown in Fig. 4.15.

4.3.6 1-jet Selection

The 1-jet bin is initially dominated by the Z/γ^* and top backgrounds (see Fig. 4.12), though the top background is efficiently reduced by vetoing events containing a b-tagged jet with $p_T > 20$ GeV (see Sect. 4.2.6).

In order to reduce the dijet background, which has a large uncertainty, two single-lepton transverse mass variables are constructed

$$m_{T,\ell_i} = \sqrt{2\, p_{T,\ell_i}\, p_T^{\text{miss,corr}} \left[1 - \cos \Delta\phi(\ell_i, p_T^{\text{miss,corr}})\right]}. \qquad (4.9)$$

In processes where $\boldsymbol{p}_T^{\text{miss,corr}}$ is collinear with a lepton, such as those with a leptonic hadron or τ decay, max $(m_{T,l})$ tends to peak at lower values (see Fig. 4.16). Therefore we require max $(m_{T,l}) > 50$ GeV in the $e\mu/\mu e$ channels. In the $ee/\mu\mu$ channels, the tighter p_T^{inv} cuts reject the dijet background so a max $(m_{T,l})$ cut is not needed.

At this stage, the $Z/\gamma^* \to \tau\tau$ background dominates the $e\mu/\mu e$ channels. To aid discrimination, a ditau mass is constructed:

$$m_{\tau\tau} = \begin{cases} m_{\ell\ell}/\sqrt{x_1 x_2} & \text{if } x_1, x_2 > 0 \\ 0 & \text{otherwise} \end{cases} \qquad (4.10)$$

where x_i is the p_T fraction of the ith τ imparted to the ith lepton. As x_i cannot be directly measured using individual neutrino momenta, $p_T^{\text{miss,corr}}$ is split by assuming the τ decays collinearly, i.e. $\Delta\phi(\ell, \nu_\tau \nu_\ell) = 0$ [24]. This approximation is reasonable since each τ has large p_T. We require $m_{\tau\tau} < m_z - 25$ GeV (see Fig. 4.16). In rare

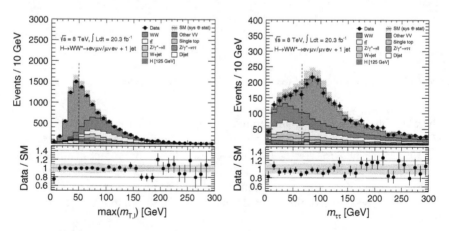

Fig. 4.16 The max $(m_{T,l})$ (*left*) and $m_{\tau\tau}$ (*right*) distributions in the $e\mu/\mu e$ channels. The selection removes events to the left of the *dashed line* in the max $(m_{T,l})$ plot, and removes events to the right of the *dashed line* in the $m_{\tau\tau}$ plot

4.3 Event Selection Criteria

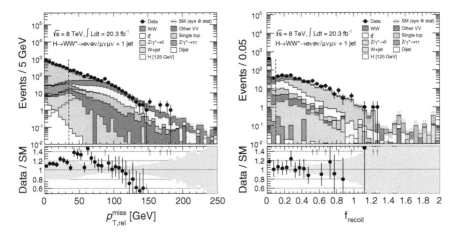

Fig. 4.17 The $p_{T,rel}^{miss}$ (*left*) and f_{recoil} (*right*) distributions in the $ee/\mu\mu$ channels. The subsequent selection removes events to the left of the *dashed line* in the $p_{T,rel}^{miss}$ plot, and removes events to the right of the *dashed line* in the f_{recoil} plot

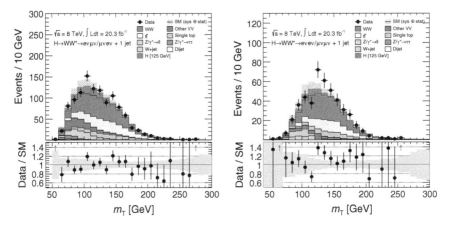

Fig. 4.18 The m_T distribution of the selected 1-jet events, in the $e\mu/\mu e$ (*left*) and $ee/\mu\mu$ (*right*) channels

cases where the ditau system is poorly reconstructed (*i.e.* $x_i < 0$), the event is accepted.

After the signal topology selection, the $Z/\gamma^* \to \ell\ell$ background is suppressed in the $ee/\mu\mu$ channels by requiring $p_{T,rel}^{miss} > 35\,\text{GeV}$ and $f_{recoil} < 0.1$ (see Fig. 4.17). In the 1-jet bin, the f_{recoil} definition is altered with respect to (4.8): the quadrant $\wedge$ is centred upon $-\boldsymbol{p}_{T,\ell\ell j}$ and the denominator becomes $p_{T,\ell\ell j}$. This is because a hard jet has been found, and so it is the dilepton + jet system that must be balanced by the soft recoil.

The m_T distributions of the selected 1-jet events are shown in Fig. 4.18.

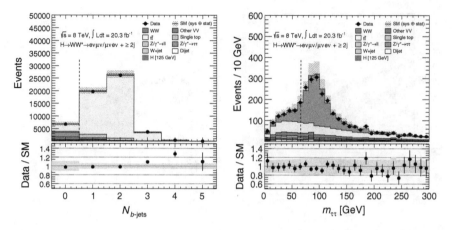

Fig. 4.19 The $N_{b\text{-jets}}$ (*left*) and $m_{\tau\tau}$ (*right*) distributions in the $e\mu/\mu e$ channels. The subsequent selection removes events to the right of each *dashed line*

4.3.7 ≥2-jet Selection

Only the $e\mu/\mu e$ channels are used in the ≥2-jet bin, which is dominated by top background (see Fig. 4.12). In the following, observable definitions using only two jets refer to the two hardest jets. As in the 1-jet bin, the top background is greatly suppressed by vetoing events featuring a b-tagged jet with $p_T > 20\,\text{GeV}$ (see Fig. 4.19). Also, the $Z/\gamma^* \to \tau\tau$ background is again suppressed by the requirement $m_{\tau\tau} < m_z - 25\,\text{GeV}$ (see Fig. 4.19).

At this point, it is necessary to contextualise the analysis. Although this thesis describes the search for the ggF production mode, searches for the other production modes (see Sect. 1.3) are also performed by the ATLAS collaboration. To simplify the statistical combination of these searches, it is helpful to avoid overlap between their signal regions. This is particularly important for the ≥2-jet bin.

The VBF production mode features two outgoing quarks at LO (see Fig. 1.3b). Therefore the ≥2-jet bin is the starting point for the VBF selection. However, the absence of colour exchange between the quarks leads to dramatically different event topologies compared to ggF events in the ≥2-jet bin; VBF events feature two high-p_T jets, separated by a large rapidity gap devoid of hadronic activity. Thus, events containing a softer jet with $p_T > 20\,\text{GeV}$ within this rapidity gap are removed from the VBF selection. This is the *central jet veto* (CJV). The VBF selection also requires that the leptons lie within this rapidity gap, since the Higgs boson is generally produced centrally. This is the *outside lepton veto* (OLV).

The VBF analysis is then split into two independent analyses: a cut-based analysis and a multivariate analysis. The cut-based VBF analysis requires a high dijet mass, $m_{jj} > 600\,\text{GeV}$, and a large rapidity gap, $\Delta y(j, j) > 3.6$. The multivariate analysis trains a boosted decision tree (BDT) [25] to discriminate VBF events based upon eight input variables:

4.3 Event Selection Criteria

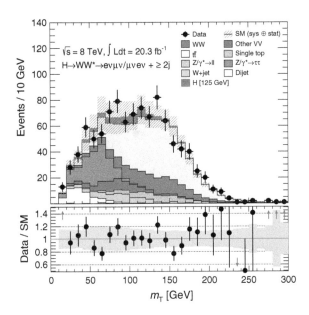

Fig. 4.20 The m_T distribution of the selected $\geq$2-jet events, in the $e\mu/\mu e$ channels

- VBF topology: m_{jj}, $\Delta y(j, j)$, η_ℓ centrality, $\sum_{\ell, j} m_{\ell j}$
- $H \to WW$ decay: $m_{\ell\ell}$, $\Delta\phi(\ell, \ell)$, m_T
- top suppression: $p_{T,\text{tot}}$

η_ℓ centrality characterises how close the leptons are to the jets. $\sum_{\ell, j} m_{\ell j}$ sums the masses of the four lepton + jet systems possible with the two hardest jets; this quantity is higher in VBF events due to the large separations involved. $p_{T,\text{tot}}$ is the magnitude of the sum of $\boldsymbol{p}_T^{\text{miss,corr}}$ and the $\boldsymbol{p}_T$ of each lepton and jet, capturing the soft hadronic recoil, which is larger in top background events. The BDT scores events between 100 % signal-like (+1) and 100 % background-like (−1), where ggF is considered a background. Events with a score greater than −0.48 are selected for the VBF analysis.

Another search considers the WH and ZH production modes (collectively known as VH), where the vector boson decays hadronically (see Fig. 1.3c). Again, the $\geq$2-jet bin is the starting point for this selection. As with VBF, the jet kinematics can be used to distinguish this from ggF. First, a small rapidity gap is required, $\Delta y(j, j) < 1.2$. Second, the dijet system must have a mass corresponding to a W or Z boson, $|m_{jj} - 85| < 15$ GeV.

To maintain orthogonality with both the VBF analyses and the VH analysis, the ggF analysis requires that events in the $\geq$2-jet bin must fail at least one cut from each selection. That is, an event must fail either the CJV, the OLV **or** the BDT score VBF cut **and** it must fail either the CJV, the OLV, the m_{jj} **or** the $\Delta y(j, j)$ VBF cut **and** it must fail either the $\Delta y(j, j)$ **or** the m_{jj} VH cut.

Finally, the usual signal topology selection is made, $m_{\ell\ell} < 55$ GeV and $\Delta\phi(\ell, \ell) < 1.8$. The m_T distribution of the selected $\geq$2-jet events is shown in Fig. 4.20.

Table 4.5 Summary of ggF event selection

	All jet bins		0-jet bin		1-jet bin		≥2-jet bin
	$e\mu/\mu e$	$ee+\mu\mu$	$e\mu/\mu e$	$ee+\mu\mu$	$e\mu/\mu e$	$ee+\mu\mu$	$e\mu/\mu e$
Pre-selection	$p_{T,\ell}^{\text{lead}} > 22$ and $p_{T,\ell}^{\text{sublead}} > 10$						
	$m_{\ell\ell} > 10$	$m_{\ell\ell} > 12$					
		$\|m_{\ell\ell} - m_Z\| > 15$					
	$p_T^{\text{miss,corr}} > 20$	$E_{T,\text{rel}}^{\text{miss}} > 40$					
Select signal	$m_{\ell\ell} < 55$						
	$\Delta\phi(\ell,\ell) < 1.8$						
Reject Z/γ^*			$p_{T,\ell\ell} > 30$	–	$m_{\tau\tau} < m_Z - 25$	–	$m_{\tau\tau} < m_Z - 25$
			–	$p_{T,\text{rel}}^{\text{miss}} > 40$	–	$p_{T,\text{rel}}^{\text{miss}} > 35$	–
			–	$f_{\text{recoil}} < 0.1$	–	$f_{\text{recoil}} < 0.1$	–
Reject fakes			$\Delta\phi(\ell\ell, p_T^{\text{inv}}) > \pi/2$		$\max(m_{T,\ell}) > 50$		–
Reject top			–	–	$N_{b\text{-jets}} = 0$	–	$N_{b\text{-jets}} = 0$
Reject VBF			–	–	–	–	Fail CJV, OLV or {BDT and $(m_{jj} > 600$ or $\Delta y(j,j) > 3.6)$}
Reject VH			–	–	–	–	Fail $\Delta y(j,j) < 1.2$ or $\|m_{jj} - 85\| < 15$

Cuts on energy, momentum and mass are given in GeV, and angular cuts are given in radians. The relevant observables are described in the text

4.3 Event Selection Criteria

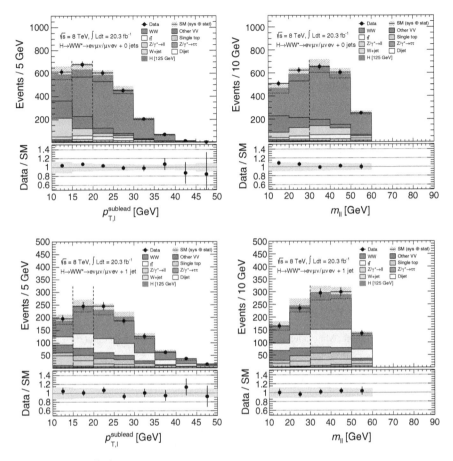

Fig. 4.21 The $p_{T,\ell}^{\text{sublead}}$ (*left*) and $m_{\ell\ell}$ (*right*) distributions in the 0-jet (*top*) and 1-jet (*bottom*) signal regions of the $e\mu/\mu e$ channels. The *dashed lines* indicate how the signal regions are split in the fit

4.3.8 Summary of Signal Regions

The entire event selection is concisely summarised in Table 4.5. The ee and $\mu\mu$ channels are combined into a single "same-flavour" channel, $ee+\mu\mu$. In total, there are 8 different signal regions: $\{e\mu, \mu e, ee+\mu\mu\} \otimes \{\text{0-jet}, \text{1-jet}\} \oplus \{e\mu, \mu e\} \otimes \{\geq\text{2-jet}\}$.

Chapter 8 shall describe how statistical limits are extracted from these signal regions, by comparing the observed m_T distributions to those expected. However, some key points are summarised here. The majority of the sensitivity lies in the 0-jet and 1-jet bins of the $e\mu/\mu e$ channels. In fact, the sensitivity is further optimised by splitting each of these four signal regions into three bins of $p_{T,\ell}^{\text{sublead}}$ and two bins of $m_{\ell\ell}$ (see Fig. 4.21). This equates to performing a three-dimensional fit of m_T, $m_{\ell\ell}$ and $p_{T,\ell}^{\text{sublead}}$. In the other signal regions, a simple one-dimensional m_T fit is used.

References

1. Particle Data Group, Review of particle physics, Phys. Rev. **D86**, 010001 (2012), and 2013 partial update for the 2014 edn
2. T. Cornelissen et al., The new ATLAS track reconstruction (NEWT). J. Phys.: Conf. Ser. **119**, 032014 (2008)
3. ATLAS Collaboration, Expected Performance of the ATLAS Experiment - Detector, Trigger and Physics, CERN-OPEN-2008-020 (2008), arXiv:0901.0512 [hep-ex]
4. G. Piacquadio, K. Prokofiev, A. Wildauer, Primary vertex reconstruction in the ATLAS experiment at LHC. J. Phys.: Conf. Ser. **119**, 032033 (2008)
5. E. Bouhova-Thacker et al., Expected Performance of Vertex Reconstruction in the ATLAS Experiment at the LHC. IEEE Trans. Nucl. Sci. **57**, 760 (2010)
6. ATLAS Collaboration, Expected electron performance in the ATLAS experiment, ATL-PHYS-PUB-2011-006 (2011)
7. ATLAS Collaboration, Improved electron reconstruction in ATLAS using the Gaussian Sum Filter-based model for bremsstrahlung, ATLAS-CONF-2012-047 (2012)
8. ATLAS Collaboration, Electron performance measurements with the ATLAS detector using the 2010 LHC proton-proton collision data, Eur. Phys. J. C **72**, 1909 (2012), arXiv:1110.3174 [hep-ex]
9. S. Laplace, J. de Vivie, Calorimeter isolation and pile-up. ATL-COM-PHYS-2012-467 (2012). (ATLAS internal)
10. ATLAS Collaboration, Electron Efficiency Measurements for 2012 and 2011 Data. ATL-COM-PHYS-2013-1287 (2013)
11. R. Nicolaidou et al., Muon identification procedure for the ATLAS detector at the LHC using Muonboy reconstruction package and tests of its performance using cosmic rays and single beam data. J. Phys.: Conf. Ser. **219**, 032052 (2010)
12. ATLAS Collaboration, Preliminary results on the muon reconstruction efficiency, momentum resolution, and momentum scale in ATLAS 2012 pp collision data. ATLAS-CONF-2013-088 (2013)
13. ATLAS Collaboration, Jet energy measurement with the ATLAS detector in proton-proton collisions at $\sqrt{s} = 7$ TeV, Eur. Phys. J. C 73, 2304 (2013), arXiv:1112.6426 [hep-ex]
14. ATLAS Collaboration, Jet energy scale and its systematic uncertainty in proton-proton collisions at $\sqrt{s} = 7$ TeV with ATLAS 2011 data, ATLAS-CONF-2013-004 (2013)
15. ATLAS Collaboration, Pile-up subtraction and suppression for jets in ATLAS, ATLAS-CONF-2013-083 (2013)
16. ATLAS Collaboration (2014), https://twiki.cern.ch/twiki/bin/view/AtlasPublic/JetEtmiss Approved2013JESUncertainty. Accessed 21 May 2014
17. ATLAS Collaboration, Selection of jets produced in proton-proton collisions with the ATLAS detector using 2011 data, ATLAS-CONF-2012-020 (2012)
18. ATLAS Collaboration, Commissioning of the ATLAS high-performance b-tagging algorithms in the 7 TeV collision data, ATLAS-CONF-2011-102 (2011)
19. ATLAS Collaboration, Calibration of b-tagging using dileptonic top pair events in a combinatorial likelihood approach with the ATLAS experiment, ATLAS-CONF-2014-004 (2014)
20. ATLAS Collaboration, Performance of missing transverse momentum reconstruction in ATLAS studied in proton-proton collisions recorded in 2012 at 8 TeV, ATLAS-CONF-2013-082 (2013)
21. ATLAS Collaboration, Measurements of the photon identification efficiency with the ATLAS detector using 4.9 fb^{-1} of pp collision data collected in 2011, ATLAS-CONF-2012-123 (2012)
22. ATLAS Collaboration, Determination of the tau energy scale and the associated systematic uncertainty in proton-proton collisions at $\sqrt{s} = 8$ TeV with the ATLAS detector at the LHC in 2012, ATLAS-CONF-2013-044 (2013)
23. ATLAS Collaboration, https://twiki.cern.ch/twiki/bin/view/AtlasPublic/MuonTriggerPublic Results. Accessed 21 May 2014

24. R.K. Ellis, I. Hinchliffe, M. Soldate, J.J. Van Der Bij, Higgs decay to $\tau^+\tau^-$: a possible signature of intermediate mass Higgs bosons at high energy hadron colliders. Nucl. Phys. B **297**, 221 (1988)
25. A. Hoecker et al., TMVA: toolkit for multivariate data analysis. PoS **ACAT**, 040 (2007), arXiv:physics/0703039

Chapter 5
Signal Modelling

This thesis describes a search for the ggF production mode (see Fig. 5.1). This exhibits large theoretical uncertainties due to higher order corrections,[1] and so its cross section is calculated at NNLO + NNLL in QCD and NLO in EW (see Fig. 1.4). PDF uncertainties are also significant, since the low-x gluon is relatively poorly constrained (see Fig. 2.2). Calculations are also sensitive to the treatment of quark masses in the loop.

Section 5.1 considers the significant theoretical issues introduced by the jet binning of the analysis, and then the MC modelling of ggF is discussed in Sect. 5.2.

5.1 Jet-Binned Cross Sections

The $gg \to H \to WW$ analysis is binned according to jet multiplicity, in order to exploit the vastly different background compositions in each jet bin (see Fig. 4.12); the 0-jet, 1-jet and $\geq$2-jet bins each have dedicated event selection criteria. Uncertainties in the expected ggF cross section must be evaluated separately for each jet bin, and correlations between these bins must be considered when they are combined. Perturbative uncertainties in the jet binning itself are considered independently from the other selection criteria, since they possess additional subtleties described below.

5.1.1 Perturbative Uncertainties in Jet-Binned Cross Sections

Consider splitting a cross section into two parts: an exclusive 0-jet cross section, σ_0, and an inclusive $\geq$1-jet cross section, $\sigma_{\geq 1}$:

---

[1] The poor convergence of the ggF perturbative series, with respect to similar $q\bar{q}$-initiated processes, is thought to be due to the larger colour factor of the gluon.

Fig. 5.1 Leading order Feynman diagram for gluon-gluon fusion (ggF)

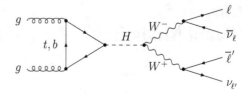

$$\sigma_{\text{tot}} = \sigma_0(p_T^{\text{cut}}) + \sigma_{\geq 1}(p_T^{\text{cut}}) \tag{5.1}$$

where p_T^{cut} is the jet p_T threshold [1]. In $\sigma_{\geq 1}$, the requirement of a jet with $p_T > p_T^{\text{cut}}$ introduces double logarithmic contributions $\alpha_S^{k+m} L^{2m}$, where $L \sim \ln(p_T^{\text{cut}}/Q)$ and Q is the scale of the hard scatter ($Q = m_H$ is typical for ggF). These terms are analogous to the logarithms introduced by soft gluon emission (see Sect. 2.1.3), though they depend upon the process and also the jet algorithm and parameters (e.g. anti-k_T with $R = 0.4$).

The schematic structures of the two inclusive cross sections are

$$\sigma_{\text{tot}} \sim \alpha_S^k \{1 \quad +\alpha_S \quad\quad\quad +\alpha_S^2 \quad\quad\quad\quad\quad\quad +(\alpha_S^3)\} \tag{5.2}$$

$$\sigma_{\geq 1} \sim \alpha_S^k \{ \quad \alpha_S(L^2+L+1) \quad +\alpha_S^2(L^4+L^3+L^2+L+1) \quad +(\alpha_S^3 L^6)\}. \tag{5.3}$$

When $p_T^{\text{cut}} \ll m_H$, the logarithms can overcome the α_S suppression and provide significant corrections to $\sigma_{\geq 1}$. When these corrections are similar in size to the perturbative corrections to σ_{tot}, the scale dependence of $\sigma_0 = \sigma_{\text{tot}} - \sigma_{\geq 1}$ is reduced by cancellations between the two series. This suggests that naïvely varying μ_R and μ_F might underestimate perturbative uncertainties. This is confirmed in Fig. 5.2, which shows that the cancellations at $p_T^{\text{cut}} = 25$ GeV (used in the $H \to WW$ analysis) are rather extreme.

When discussing uncertainties in jet-binned cross sections, it is convenient to consider a general parametrisation of the covariance matrix [2]. In the $\{\sigma_0, \sigma_{\geq 1}\}$ basis, the covariance matrix is decomposed into two uncertainty sources

$$C = C^{\text{yield}} + C^{\text{migration}} \tag{5.4}$$

$$= \begin{pmatrix} (\Delta_0^y)^2 & \Delta_0^y \Delta_{\geq 1}^y \\ \Delta_0^y \Delta_{\geq 1}^y & (\Delta_{\geq 1}^y)^2 \end{pmatrix} + \left(\Delta_{0\to}^{\text{mig}}\right)^2 \begin{pmatrix} +1 & -1 \\ -1 & +1 \end{pmatrix} \tag{5.5}$$

where Δ_N^y is the uncertainty in the N-jet cross section due to the yield uncertainty, and $\Delta_{0\to}^{\text{mig}}$ is the cross section uncertainty due to bin migrations. The yield component is fully correlated between jet bins, though can affect each with different magnitudes. The migration component is fully anti-correlated and affects both bins

5.1 Jet-Binned Cross Sections

equally, conserving the normalisation. In the statistical model (see Sect. 8.2.2), these two sources are treated as nuisance parameters θ with uncertainty amplitudes $\nu(\Delta\theta)$ in the $\{\sigma_0, \sigma_{\geq 1}\}$ basis:

$$\begin{aligned} \theta^{\text{yield}} &: \left(\Delta_0^y, \quad \Delta_{\geq 1}^y \right) \\ \theta_{0\rightarrow}^{\text{mig}} &: \left(\Delta_{0\rightarrow}^{\text{mig}}, \quad -\Delta_{0\rightarrow}^{\text{mig}} \right). \end{aligned} \quad (5.6)$$

Different prescriptions for evaluating perturbative uncertainties are defined by their choice of Δ_0^y, $\Delta_{\geq 1}^y$ and $\Delta_{0\rightarrow}^{\text{mig}}$. This includes a choice of which observables to measure uncertainties in, and also the method of measuring the uncertainties. For example, the naïve prescription described above is equivalent to choosing

$$\text{Naïve:} \quad \Delta_0^y = \Delta\sigma_0, \qquad \Delta_{\geq 1}^y = \Delta\sigma_{\geq 1}, \qquad \Delta_{0\rightarrow}^{\text{mig}} = 0 \quad (5.7)$$

where uncertainties are evaluated at fixed order by varying μ_R and μ_F.

In the $H \rightarrow WW$ analysis, there is also an exclusive 1-jet bin. The second jet veto introduces an additional source of migrations, now between the 1-jet and $\geq$2-jet bins. Therefore, in the $\{\sigma_0, \sigma_1, \sigma_{\geq 2}\}$ basis, the covariance matrix has three components

$$C = \begin{pmatrix} (\Delta_0^y)^2 & \Delta_0^y\Delta_1^y & \Delta_0^y\Delta_{\geq 2}^y \\ \Delta_0^y\Delta_1^y & (\Delta_1^y)^2 & \Delta_1^y\Delta_{\geq 2}^y \\ \Delta_0^y\Delta_{\geq 2}^y & \Delta_1^y\Delta_{\geq 2}^y & (\Delta_{\geq 2}^y)^2 \end{pmatrix} \\ + \left(\Delta_{0\rightarrow}^{\text{mig}}\right)^2 \begin{pmatrix} +1 & -(1-\rho) & -\rho \\ -(1-\rho) & (1-\rho)^2 & \rho(1-\rho) \\ -\rho & \rho(1-\rho) & \rho^2 \end{pmatrix} + \left(\Delta_{1\rightarrow}^{\text{mig}}\right)^2 \begin{pmatrix} 0 & 0 & 0 \\ 0 & +1 & -1 \\ 0 & -1 & +1 \end{pmatrix} \quad (5.8)$$

where ρ is the fraction of migrations from the 0-jet bin ($\Delta_{0\rightarrow}^{\text{mig}}$) that enter the $\geq$2-jet bin. In this case, there are three nuisance parameters θ with uncertainty amplitudes $\nu(\Delta\theta)$ in the $\{\sigma_0, \sigma_1, \sigma_{\geq 2}\}$ basis:

$$\begin{aligned} \theta^{\text{yield}} &: \left(\Delta_0^y, \quad \Delta_1^y, \quad \Delta_{\geq 2}^y \right) \\ \theta_{0\rightarrow}^{\text{mig}} &: \left(\Delta_{0\rightarrow}^{\text{mig}}, \quad -(1-\rho)\Delta_{0\rightarrow}^{\text{mig}}, \quad -\rho\Delta_{0\rightarrow}^{\text{mig}} \right) \\ \theta_{1\rightarrow}^{\text{mig}} &: \left(0, \quad \Delta_{1\rightarrow}^{\text{mig}}, \quad -\Delta_{1\rightarrow}^{\text{mig}} \right). \end{aligned} \quad (5.9)$$

So it is Δ_0^y, Δ_1^y, $\Delta_{\geq 2}^y$, $\Delta_{0\rightarrow}^{\text{mig}}$, $\Delta_{1\rightarrow}^{\text{mig}}$ and ρ that must be determined. Two different prescriptions for evaluating these shall now be examined.

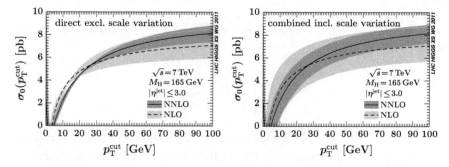

Fig. 5.2 The exclusive 0-jet ggF cross section versus the jet p_T threshold [1]. The bands show the perturbative uncertainties evaluated using the naïve prescription (*left*) and the combined inclusive prescription (*right*)

5.1.2 Combined Inclusive Prescription

The *combined inclusive* (CI) prescription[2] [3] uses scale variations of $\sigma_{\geq 1}$ and $\sigma_{\geq 2}$ to probe the size of the higher order logarithmic corrections, and uses these to estimate the bin migration uncertainties. It therefore chooses

$$\text{CI:} \quad \Delta_0^y = \Delta\sigma_{\text{tot}}, \qquad \Delta_1^y = 0, \qquad \Delta_{\geq 2}^y = 0,$$
$$\Delta_{0\to}^{\text{mig}} = \Delta\sigma_{\geq 1}, \qquad \Delta_{1\to}^{\text{mig}} = \Delta\sigma_{\geq 2}, \qquad \rho = 0 \qquad (5.10)$$

where $\Delta\sigma_{\text{tot}}$, $\Delta\sigma_{\geq 1}$ and $\Delta\sigma_{\geq 2}$ are evaluated, at a common fixed order in α_S, through variations of the μ_R and μ_F scales.

This is equivalent to assuming inclusive cross sections have uncorrelated uncertainties

$$\text{CI:} \quad \sigma_N = \sigma_{\geq N} - \sigma_{\geq N+1} \quad \Rightarrow \quad \Delta\sigma_N^2 = \Delta\sigma_{\geq N}^2 + \Delta\sigma_{\geq N+1}^2. \qquad (5.11)$$

Although this assumption is not believed to be exact, the CI prescription offers a practical solution to the cancellations described in Sect. 5.1.1 (see Fig. 5.2). It ensures that uncertainties in exclusive cross sections are larger than those in the corresponding inclusive cross section, i.e. $\Delta\sigma_N \geq \Delta\sigma_{\geq N}$, with the equality valid in the large-p_T^{cut} limit.

It should be emphasised that each inclusive cross section must be evaluated at the same order in α_S (e.g. $\sigma_{\text{tot}}^{\text{NNLO}}$, $\sigma_{\geq 1}^{\text{NLO}}$ and $\sigma_{\geq 2}^{\text{LO}}$). This can restrict the application of the CI prescription. For example, an exclusive 2-jet bin cannot be added until σ_{tot} is calculated at N^3LO in QCD. For the ggF contamination to the VBF signal

---

[2] The CI prescription is also called the Stewart-Tackmann prescription, after its original proponents.

5.1 Jet-Binned Cross Sections

region (defined with a central jet veto), $\sigma_{\geq 2}^{\text{NLO}}$ and $\sigma_{\geq 3}^{\text{LO}}$ can be used since they are evaluated in a significantly different phase space ($\Delta\sigma_{\geq 2}^{\text{NLO}}$ and $\Delta\sigma_{\geq 2}^{\text{LO}}$ are treated as fully correlated).

HNNLO [4] is used to compute $\sigma_{\text{tot}}^{\text{NNLO}}$, $\sigma_{\geq 1}^{\text{NLO}}$ and $\sigma_{\geq 2}^{\text{LO}}$. However, the CI prescription can be improved by using the NNLO + NNLL(QCD)+NLO(EW) σ_{tot} calculation [5], which has smaller perturbative uncertainties than $\sigma_{\text{tot}}^{\text{NNLO}}$. In order to combine these results whilst preserving the total normalisation, the jet bin fractions $f_N = \sigma_N/\sigma_{\text{tot}}$ from HNNLO are used to propagate the $\Delta\sigma_{\geq N}$ to $\Delta\sigma_N$:

$$\delta\sigma_0^2 = \frac{1}{f_0^2}\delta\sigma_{\text{tot}}^2 + \left(\frac{1}{f_0} - 1\right)^2 \delta\sigma_{\geq 1}^2 \tag{5.12}$$

$$\delta\sigma_1^2 = \left(\frac{1-f_0}{f_1}\right)^2 \delta\sigma_{\geq 1}^2 + \left(\frac{1-f_0}{f_1} - 1\right)^2 \delta\sigma_{\geq 2}^2 \tag{5.13}$$

where $\delta\sigma_i = \Delta\sigma_i/\sigma_i$. This assumes the uncertainties are Gaussian distributed, though the nuisance parameters are constrained by log-normal distributions in the statistical model (see Sect. 8.2.2). The $\Delta\sigma_{\geq N}$ are evaluated via independent variation of μ_R and μ_F in the range $m_H/4 \leq \mu_R, \mu_F \leq m_H$, whilst observing the constraint $1/2 \leq \mu_R/\mu_F \leq 2$. These are then propagated to $\Delta\sigma_N$ using (5.12) and (5.13), as shown in Table 5.1.

5.1.3 Jet Veto Efficiency Prescription

The *jet veto efficiency* (JVE) prescription [6, 7] considers each exclusive cross section as a product of the total cross section and jet veto efficiencies

Table 5.1 Inputs and outputs *(boxed)* of the combined inclusive prescription for ggF, with $m_H = 125\,\text{GeV}$ and $\sqrt{s} = 8\,\text{TeV}$

i	$f_i = \sigma_i/\sigma_{\text{tot}}$	σ_i (pb)	$\Delta\sigma_i/\sigma_i$
≥ 0	–	19.27	7.8%
≥ 1	–	–	20.2%
≥ 2	–	–	69.7%
0	0.614	11.83	18.0%
1	0.267	5.15	42.6%
≥ 2	–	2.29	69.7%

All inputs are computed by HNNLO except σ_{tot} and $\Delta\sigma_{\text{tot}}$, which are from reference [5]

$$\sigma_0 = \sigma_{tot}\epsilon_0 \tag{5.14}$$

$$\sigma_1 = \sigma_{tot}(1 - \epsilon_0)\epsilon_1 \tag{5.15}$$

$$\sigma_{\geq 2} = \sigma_{tot}(1 - \epsilon_0)(1 - \epsilon_1) \tag{5.16}$$

where ϵ_0 and ϵ_1 are the first and second jet veto efficiencies. That is, $\epsilon_0 = \epsilon_0\left(p_T^{cut}\right)$ is the efficiency of a veto upon jets with $p_T > p_T^{cut}$, and $\epsilon_1 = \epsilon_1\left(p_T^{sel}, p_T^{cut}\right)$ is the efficiency of a veto upon additional jets with $p_T > p_T^{cut}$ given that there is already a jet with $p_T > p_T^{sel}$. In the $H \to WW$ analysis, $p_T^{cut} = p_T^{sel} = 25$ GeV.[3]

The JVE prescription assumes that uncertainties in σ_{tot} and the ϵ_N are uncorrelated, which ensures that uncertainties in exclusive cross sections are larger than those in the total cross section, i.e. $\Delta\sigma_N > \Delta\sigma_{tot}$. This is equivalent to choosing

JVE: $\Delta_0^y = \Delta\sigma_{tot}\epsilon_0,$ $\Delta_1^y = \Delta\sigma_{tot}(1 - \epsilon_0)\epsilon_1,$ $\Delta_{\geq 2}^y = \Delta\sigma_{tot}(1 - \epsilon_0)(1 - \epsilon_1),$

$\Delta_{0\to}^{mig} = \Delta\epsilon_0\sigma_{tot},$ $\Delta_{1\to}^{mig} = \Delta\epsilon_1\sigma_{tot}(1 - \epsilon_0),$ $\rho = 1 - \epsilon_1.$ (5.17)

Thus, the JVE prescription requires six inputs: $\sigma_{tot}, \epsilon_0, \epsilon_1, \Delta\sigma_{tot}, \Delta\epsilon_0$ and $\Delta\epsilon_1$. As in the CI prescription, σ_{tot} and $\Delta\sigma_{tot}$ are taken from the NNLO + NNLL(QCD) + NLO (EW) calculation. However, the ϵ_N contain similar cancellations to those discussed in Sect. 5.1.1, and so the $\Delta\epsilon_N$ must be treated with care.

There is an ambiguity in the definition of ϵ_N that is not present in the fixed order $\sigma_{\geq N}$ calculations [6]. For example, considering NNLO terms with respect to the N-jet process, three alternative definitions of ϵ_N may be identified

$$\epsilon_N^{(a)} = 1 - \frac{\sigma_{\geq N+1}^{NLO}}{\sigma_{\geq N}^{NNLO}} \tag{5.18}$$

$$\epsilon_N^{(b)} = 1 - \frac{\sigma_{\geq N+1}^{NLO}}{\sigma_{\geq N}^{NLO}} \tag{5.19}$$

$$\epsilon_N^{(c)} = 1 - \frac{\sigma_{\geq N+1}^{NLO}}{\sigma_{\geq N}^{LO}} + \left(\frac{\sigma_{\geq N}^{NLO}}{\sigma_{\geq N}^{LO}} - 1\right)\frac{\sigma_{\geq N+1}^{LO}}{\sigma_{\geq N}^{LO}}. \tag{5.20}$$

Although scheme (a) is the most intuitive definition, schemes (b) and (c) differ by N^3LO terms and therefore probe higher order corrections.[4] However, this probing is less susceptible to accidental cancellations than scale variations. Thus, $\epsilon_N^{(a)}$ is used as the nominal ϵ_N, whilst $\Delta\epsilon_N$ is evaluated by scale variations in $\epsilon_N^{(a)}$ or by the difference between the schemes, whichever is larger.

Figure 5.3 shows how scheme differences of ϵ_0 inflate the perturbative uncertainties compared to scale variations of $\epsilon_0^{(a)}$. At $p_T^{cut} = 25$ GeV, it increases $\Delta\epsilon_0$ from ~ 5

---

[3] Since the jets are predominantly central, the raising of p_T^{cut} to 30 GeV in the forward region is neglected. This approximation is conservative, since the $\Delta\sigma_N$ decrease with higher p_T^{cut}.
[4] Processes whose perturbative series converge more quickly, such as $q\bar{q} \to Z$, exhibit better agreement between schemes. This supports using these schemes to evaluate perturbative uncertainties.

5.1 Jet-Binned Cross Sections

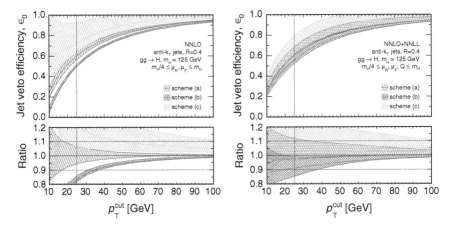

Fig. 5.3 Jet veto efficiency ϵ_0 versus the jet p_T threshold, computed with fixed order (*left*) and resummed (*right*) calculations by JETVHETO [7]. The bands show scale uncertainties. The bands of schemes (b) and (c) are not used in $\Delta\epsilon_0$

to ~20%. Figure 5.3 also shows how resummation of the $\ln\left(p_T^{\text{cut}}/m_H\right)$ logarithms to all orders of α_S can improve the estimation of ϵ_0, resulting in better agreement between schemes and consequently reducing $\Delta\epsilon_0$. This resummation includes NNLL terms and is performed by JETVHETO [7].

The three NNLO schemes (5.18)–(5.20) can also be used to define ϵ_1. This offers an improvement compared to the CI prescription, which is currently limited to using $\sigma_{\geq 1}^{\text{NLO}}$ and $\sigma_{\geq 2}^{\text{LO}}$ (see Sect. 5.1.2). Unfortunately, it is not possible to calculate $\epsilon_1^{(a)}$ since a full $\sigma_{\geq 1}^{\text{NNLO}}$ calculation is not yet available. Instead, we choose $\epsilon_1 = (\epsilon_1^{(b)} + \epsilon_1^{(c)})/2$ and $\Delta\epsilon_1$ is evaluated by an envelope of scale uncertainties in both $\epsilon_1^{(b)}$ and $\epsilon_1^{(c)}$, which are calculated using MCFM [8]. The validity of this approximation is tested using gg-initiated diagrams only, for which a $\sigma_{\geq 1}^{\text{NNLO}}$ calculation exists [9]. For k_T jets with $R = 0.5$ and $p_T^{\text{cut}} = 30\,\text{GeV}$, we find that $\epsilon_1^{(a)} = 0.831$, $\epsilon_1^{(b)} = 0.761$ and $\epsilon_1^{(c)} = 0.843$.

The NNLO + NNLL(QCD) + NLO(EW) σ_{tot} calculation, the JETVHETO NNLO + NNLL ϵ_0 calculation and the MCFM fixed order ϵ_1 calculation are used as inputs to the JVE prescription (5.17). Table 5.2 shows the jet-binned cross sections and uncertainties obtained using simple Gaussian propagation of uncertainties, though the nuisance parameters are constrained by log-normal distributions in the statistical model (see Sect. 8.2.2).

Figure 5.4 compares the above ϵ_0 and ϵ_1 calculations to a variety of different POWHEGBOX + PYTHIA8 configurations. The ϵ_N calculations are performed at parton-level in the large-m_t limit; the solid lines show similar configurations of POWHEGBOX + PYTHIA8, and can be considered directly comparable. The effect of consecutively adding hadronisation, MPI and finite quark mass effects are also shown. Finally, the blue line shows how reweighting the Higgs boson p_T distribution

Table 5.2 Results of the jet veto efficiency prescription for ggF, with $m_H = 125\,\text{GeV}$ and $\sqrt{s} = 8\,\text{TeV}$

	x	$\Delta x/x$
σ_{tot} (pb)	19.27 ± 1.50	7.8 %
ϵ_0	0.613 ± 0.072	11.7 %
ϵ_1	0.615 ± 0.061	9.9 %
σ_0 (pb)	11.81 ± 1.66	14.1 %
σ_1 (pb)	4.59 ± 1.03	22.4 %
$\sigma_{\geq 2}$ (pb)	2.87 ± 0.73	25.4 %

σ_{tot} is from reference [5], ϵ_0 is from JETVHETO and ϵ_1 is from MCFM

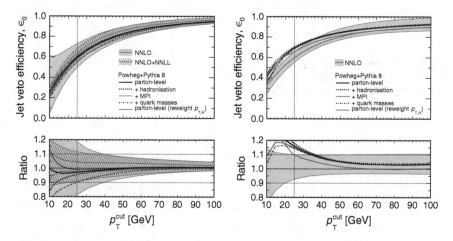

Fig. 5.4 Veto efficiencies of a first jet (*left*) and a second jet (*right*), versus the jet p_T threshold. Fixed order (*green*) and resummed (*red*) results are shown with their perturbative uncertainties, and are compared to different MC scenarios

(see Sect. 5.2.1) influences these observables. Figure 5.4 shows that, following $p_{T,H}$ reweighting, the MC is within the perturbative uncertainty band for both ϵ_0 and ϵ_1.

5.1.4 Discussion of Results

It is helpful to directly compare the predicted jet-binned cross sections of the two prescriptions, as in Fig. 5.5. "Fixed order CI" and "resummed JVE" are the prescriptions described in the preceding sections. JVE offers a reduction in uncertainty compared to CI, whilst both prescriptions remain consistent with the POWHEGBOX + PYTHIA8 MC used, both before and after $p_{T,H}$ reweighting. The improvement is due to the resummation of large logarithms in ϵ_0 and including higher order terms in ϵ_1. The JVE prescription is chosen to estimate the perturbative uncertainties in the jet binning.

5.1 Jet-Binned Cross Sections

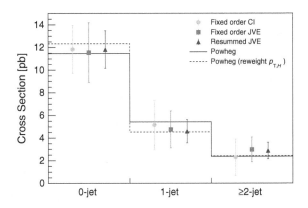

Fig. 5.5 Jet-binned cross sections for ggF production with $m_H = 125\,\text{GeV}$ and $\sqrt{s} = 8\,\text{TeV}$. The error bars show the perturbative uncertainty associated with each prescription. Consistency with POWHEGBOX + PYTHIA8 is also shown, normalised to 19.27 pb

"Fixed order JVE" replaces the resummed ϵ_0 calculation with the fixed order ϵ_0 calculation (i.e. replace the red band with the green band in Fig. 5.4). This has the same formal accuracy as "fixed order CI" when calculating σ_0, and so their comparison in the 0-jet bin directly evaluates how conservative each prescription is. JVE is found to be more conservative than CI.

Recently, a more general prescription for evaluating perturbative uncertainties in jet-binned cross sections has been proposed, to directly estimate each uncertainty amplitude in (5.9) [2]. Although it offers a further reduction in uncertainty, this prescription is not considered here. It is not viable as it predicts σ_{tot} to be $\sim 10\,\%$ larger than that presented in reference [5], which is universally used by the ATLAS and CMS collaborations for analyses of the Run I dataset. Additionally, it includes incomplete higher order terms in the perturbative series, and employs a controversial π^2 resummation technique.

5.2 Monte Carlo Modelling

The ggF signal is modelled by POWHEGBOX + PYTHIA8, including the exact mass dependence of the t and b quarks in the loop [10], and using the CT10 PDF [11] to describe the incoming partons. Various aspects of this modelling are now discussed.

5.2.1 Higgs Boson Transverse Momentum

The MC events are produced with the POWHEGBOX parameter `hfact` tuned to $m_H/1.2$.[5] This setting reproduces the NNLO + NNLL $p_{T,H}$ distribution calculated

---

[5] `hfact` controls the scale at which the first emission transitions from Sudakov-like to ME-like. Its use in tuning $p_{T,H}$ is discussed in Sects. 4.5 and 4.9 of reference [1].

by HQT 2.0 [12] in the large-m_t limit, when effects of hadronisation, MPI and finite quark masses are turned off in POWHEGBOX + PYTHIA8.

The $p_{T,H}$ distribution of the generated MC events is reweighted to the best available prediction.[6] Simply reweighting the inclusive $p_{T,H}$ distribution is found to underestimate $\sigma_{\geq 2}$ (cf. Sect. 5.1), due to correlations between $p_{T,H}$ and N_{jets}. Thus, the reweighting is required to simultaneously satisfy three criteria:

- the inclusive $p_{T,H}$ distribution agrees with HRES 2.1 [13],
- the $p_{T,H}$ distribution in the $\geq$2-jet bin agrees with the $gg \to Hjj$ process calculated by MINLO + PYTHIA8 [14],
- and the jet-binned cross sections agree with Sect. 5.1.

HRES 2.1 computes the inclusive $p_{T,H}$ distribution with NNLO + NNLL accuracy and includes finite m_t and m_b effects in the loop. It also employs a dynamic scale of $\mu_0 = \sqrt{m_H^2 + p_{T,H}^2}$ as the nominal μ_R and μ_F scale, which improves results at high $p_{T,H}$ compared to the fixed $\mu_0 = m_H$ scale used by HQT 2.0. MINLO is an improved version of POWHEGBOX, which includes higher order logarithmic contributions through the careful choice of μ_R and μ_F scales [14]. Figure 5.5 shows that the reweighting preserves agreement with the predicted N_{jets} distribution.

5.2.2 Event Selection Acceptance

In order to measure the total ggF cross section, the signal acceptance of the object and event selections must be estimated. The extrapolation from the measured phase space to the inclusive phase space introduces uncertainties. It is helpful to separate theoretical uncertainties from the others by measuring an intermediate cross section in a *fiducial region* of phase space, as defined in Table 5.3 for each signal region. Events featuring leptonic τ decays are excluded from the fiducial region, in order to make it easier for theorists to calculate the expected fiducial cross section within their model of choice. These criteria use hadron-level objects and are chosen to closely represent those of the detector-level selection, in order to minimise the extrapolation to the fiducial region.

To define the hadron-level objects, the MC event record is used to identify prompt charged leptons and neutrinos. A $E_{T,\nu\nu}$ vector is constructed from the neutrinos. Each charged lepton is 'dressed' by adding the four-momenta of photons within a cone of $\Delta R < 0.1$, in order to recover energy lost via QED FSR. Jets are found using individual particles as inputs (cf. topo-clusters at detector-level). Muons and neutrinos are excluded from jet finding since they interact weakly with the calorimeter. Objects must pass the same p_T, η and overlap removal criteria applied at detector-level.

---

[6] This $p_{T,H}$ reweighting is motivated by the $H \to \gamma\gamma$ and $H \to \tau\tau$ analyses, which feature boosted Higgs boson selection categories.

5.2 Monte Carlo Modelling

Table 5.3 Hadron-level event selection criteria for each fiducial region

Jet binning	$e\mu/\mu e$	$ee/\mu\mu$
Inclusive	$p_{T,\ell}^{\text{lead}} > 22$ and $p_{T,\ell}^{\text{sublead}} > 10$	
	$m_{\ell\ell} > 10$	$m_{\ell\ell} > 12$
	–	$\|m_{\ell\ell} - m_Z\| > 15$
	$p_{T,vv} > 20$	$E_{T,\text{rel},vv} > 40$
	$m_{\ell\ell} < 55$	
	$\Delta\phi(\ell,\ell) < 1.8$	
0-jet	$\Delta\phi(\ell\ell, p_T^{\text{inv}}) > \pi/2$	
	$p_{T,\ell\ell} > 30$	
1-jet	$\max(m_{T,\ell}) > 50$	–
	$m_{\tau\tau} < m_Z - 25$	–
$\geq$2-jet	$m_{\tau\tau} < m_Z - 25$	–
	Fail $\Delta y(j,j) > 3.6$ or $m_{jj} > 600$ or CJV or OLV	

Events featuring leptonic τ decays are excluded from the fiducial region. Cuts on energy, momentum and mass are given in GeV, and angular cuts are given in radians. The CJV and OLV are the central jet veto and outside lepton veto, respectively. See Chap. 4 for a detailed explanation of the criteria

The measured fiducial cross section is extracted by

$$\sigma_{\text{ggF}}^{\text{fid}} = \frac{N_{\text{obs}} - N_{\text{bkg}}}{C_{\text{ggF}} \cdot L} \tag{5.21}$$

where N_{obs} is the observed number of events, N_{bkg} is the expected number of background events, L is the luminosity, and C_{ggF} is the ratio of the expected number of ggF events passing the detector-level selection to those passing the fiducial selection. C_{ggF} accounts for detector effects such as lepton trigger and reconstruction efficiencies and object mismeasurement due to the finite resolution of the detector.

On the other hand, the measured total cross section is extracted by

$$\sigma_{\text{ggF}} = \frac{N_{\text{obs}} - N_{\text{bkg}}}{C_{\text{ggF}} \cdot A_{\text{ggF}} \cdot \text{BR} \cdot L} \tag{5.22}$$

where A_{ggF} is the ratio of the expected number of ggF events passing the fiducial selection to the total expected number of ggF events. A_{ggF} accounts for the acceptance of the event selection criteria. BR incorporates the branching ratios of the Higgs and W bosons for the channel in question.

Fiducial cross sections are only extracted from the 0-jet and 1-jet bins of the $e\mu/\mu e$ channels, since these are the most sensitive signal regions. The acceptances C_{ggF}, A_{ggF} and $C_{\text{ggF}} \cdot A_{\text{ggF}}$ are displayed in Table 5.4, together with their respective uncertainties.

Theoretical uncertainties in the acceptance (other than the jet binning uncertainties discussed in Sect. 5.1) are evaluated at hadron-level by changing some aspect of the MC modelling and measuring the change in acceptance relative to the jet-binned

Table 5.4 The signal acceptances C_{ggF}, A_{ggF} and $C_{\text{ggF}} \cdot A_{\text{ggF}}$

	0-jet	1-jet
C_{ggF} ($\times 100$)	50.7 ± 2.7	50.6 ± 2.2
Trigger efficiency	0.7 %	0.6 %
Lepton efficiency	2.6 %	2.4 %
Lepton p_T scale and resolution	1.1 %	0.9 %
Jet energy scale and resolution	4.3 %	2.2 %
Jet b-tagging efficiency	0.0 %	1.8 %
p_T^{inv} modelling	0.1 %	0.1 %
PS/UE	1.5 %	2.0 %
A_{ggF} ($\times 100$)	20.6 ± 3.0	7.5 ± 1.7
Perturbative QCD		
Jet binning	14 %	22 %
Other cuts	1.1 %	1.4 %
PDFs	3.7 %	3.4 %
PS/UE	2.2 %	3.8 %
NLO-PS	1.8 %	1.1 %
$C_{\text{ggF}} \cdot A_{\text{ggF}}$ ($\times 100$)	10.4 ± 1.6	3.8 ± 0.9

A breakdown of the relative uncertainties from different sources is also shown

cross sections. In the case of PDF uncertainties, the acceptance is calculated relative to the total cross section, in order to include PDF uncertainties in the jet binning. Since these uncertainties are used in extracting the signal strength, the fiducial volume is modified to include events featuring lepton τ decays.

Four sources of theoretical uncertainty are considered:

- higher order corrections,
- PDFs,
- parton shower, hadronisation and underlying event models,
- NLO-PS matching scheme.

Uncertainties due to higher order corrections are evaluated via independent variation of renormalisation and factorisation scales in the range $m_H/2 \leq \mu_R, \mu_F \leq 2m_H$, whilst observing the constraint $1/2 \leq \mu_R/\mu_F \leq 2$. In the ≥ 2-jet bin, this is evaluated using NLO MC of the $gg \to H + 1$ jet process, since the NLO MC of the inclusive $gg \to H$ process relies upon the parton shower to model the second jet.

Uncertainties due to PDFs are evaluated in two ways. First, the acceptance is compared to that predicted with the MSTW2008 PDF [15]. Second, the set of PDF eigenvectors corresponding to 90 % CL of the CT10 fit were used to evaluate an uncertainty, which was then rescaled to 68 % CL (assuming a Gaussian distribution). PDF uncertainties are evaluated using MC@NLO.

Uncertainties due to the parton shower (PS), hadronisation and underlying event (UE) models are evaluated by comparing POWHEGBOX showered by PYTHIA8 (nom-

5.2 Monte Carlo Modelling

Table 5.5 Theoretical uncertainties in the ggF acceptance for each signal region used in the fitting procedure

	$m_{\ell\ell}$ (GeV)	$p_{T,\ell}^{sublead}$ (GeV)	QCD scale (%)	PDF (%) Collab.	PDF (%) 68 % CL	PS/UE (%) PYTHIA6	PS/UE (%) HERWIG	NLO-PS (%)
$ee/\mu\mu$ channels								
0-jet	12–55	>10	1.4	+1.9	3.2	+1.6	+6.4	−2.5
1-jet	12–55	>10	1.9	+1.8	2.8	(−)1.5	+2.1	(−)1.4
$e\mu/\mu e$ channels								
0-jet	10–30	10–15	2.6	+1.8	3.2	−1.7	+5.7	−3.5
		15–20	1.3	+1.9	3.2	(+)2.4	+4.9	−2.9
		>20	1.0	+1.9	3.2	−2.2	(−)1.6	(−)1.4
	30–55	10–15	1.5	+1.8	3.3	(+)2.0	+5.5	−3.8
		15–20	1.5	+1.9	3.3	(−)2.5	(+)2.4	−2.5
		>20	3.5	+1.9	3.3	−1.9	−2.4	(−)1.3
1-jet	10–30	10–15	3.2	+1.7	2.9	+2.9	+10.8	−3.8
		15–20	2.9	+1.8	2.9	(+)3.8	(+)3.9	(+)3.6
		>20	3.5	+1.8	2.7	(+)2.1	(+)2.0	(−)1.9
	30–55	10–15	5.8	+1.7	3.0	(+)3.2	+11.4	−6.8
		15–20	1.0	+1.8	3.3	(+)2.6	+13.5	+6.7
		>20	1.3	+1.8	2.8	(−)1.9	(−)1.8	(+)1.7
≥2-jet	10–55	>10	3.6	+2.0	2.2	(−)1.7	(+)1.7	−4.5

PDF uncertainties are in acceptances relative to the inclusive cross section, whereas others are calculated within jet bins. When the uncertainty is statistically insignificant, the statistical uncertainty on the generator difference is given, and the sign of the generator difference is parenthesised. These are the uncertainties used in the fit

inal), PYTHIA6 and HERWIG. Uncertainties due to the NLO-PS matching scheme are evaluated by comparing POWHEGBOX + HERWIG to MC@NLO + HERWIG++.

Theoretical acceptance uncertainties are also calculated for every signal region used in the fitting procedure, including the individual signal regions split by $p_{T,l}^{sublead}$ and $m_{\ell\ell}$. These uncertainties are shown in Table 5.5, and are evaluated in the fiducial regions described in Table 5.3.

5.2.3 m_T Shape Modelling

Theoretical uncertainties in the shape of the m_T distribution are also investigated. Uncertainties due to scale, PS/UE and NLO-PS choices are considered using the methods described above (PDF uncertainties are neglected). The split signal regions are not used in this study since the statistical fluctuations in the m_T distributions are large.

Each uncertainty is parametrised by fitting the ratio of the m_T shapes, and then symmetrising the fit to produce "up" and "down" variations. The m_T distributions are

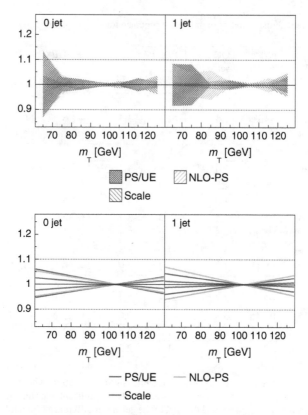

Fig. 5.6 ggF m_T shape systematic uncertainties in the 0-jet and 1-jet signal regions of the $e\mu/\mu e$ channels. The raw uncertainties are shown above and the fits are shown below. The constant (flat) terms of the fit are outside the visible x-range

normalised to unit integral in order to remove effects from acceptance uncertainties. In cases where multiple variations exist within a single uncertainty source (such as the six scale variations), the largest deviation from the nominal result is fit. A linear fit is used in the central m_T region, and a constant is used in the low-m_T and high-m_T tails of the distribution where statistical fluctuations dominate.

These fits allow the hadron-level m_T distribution of the ggF signal to be reweighted to the "up" and "down" variations. In this way, the m_T shape uncertainty is treated as a nuisance parameter in the $H \rightarrow WW$ fitting procedure. The uncertainties for the 0-jet and 1-jet signal regions are displayed in Fig. 5.6.

References

1. LHC Higgs Cross Section Working Group, Handbook of LHC Higgs Cross Sections: 2. Differential Distributions, CERN-2012-002 (2012). arXiv:1201.3084 [hep-ph]
2. R. Boughezal, X. Liu, F. Petriello, F.J. Tackmann, J.R. Walsh, Combining resummed Higgs predictions across jet bins. Phys. Rev. D **89**, 074044 (2014). arXiv:1312.4535 [hep-ph]
3. I.W. Stewart, F.J. Tackmann, Theory uncertainties for Higgs and other searches using jet bins. Phys. Rev. D **85**, 034011 (2012). arXiv:1107.2117 [hep-ph]
4. M. Grazzini, NNLO predictions for the Higgs boson signal in the $H \to WW \to l\nu l\nu$ and $H \to ZZ \to 4l$ decay channels. JHEP **0802**, 043 (2008). arXiv:0801.3232 [hep-ph]
5. LHC Higgs Cross Section Working Group, Handbook of LHC Higgs Cross Sections: 3. Higgs Properties, CERN-2013-004 (2013). arXiv:1307.1347 [hep-ph]
6. A. Banfi, G.P. Salam, G. Zanderighi, NLL+NNLO predictions for jet-veto efficiencies in Higgs-boson and Drell-Yan production. JHEP **1206**, 159 (2012). arXiv:1203.5773 [hep-ph]
7. A. Banfi, P.F. Monni, G.P. Salam, G. Zanderighi, Higgs and Z-boson production with a jet veto. Phys. Rev. Lett. **109**, 202001 (2012). arXiv:1206.4998 [hep-ph]
8. J.M. Campbell, R.K. Ellis, C. Williams, Hadronic production of a Higgs boson and two jets at next-to-leading order. Phys. Rev. D **81**, 074023 (2010). arXiv:1001.4495 [hep-ph]
9. R. Boughezal, F. Caola, K. Melnikov, F. Petriello, M. Schulze, Higgs boson production in association with a jet at next-to-next-to-leading order in perturbative QCD. JHEP **1306**, 072 (2013). arXiv:1302.6216 [hep-ph]
10. E. Bagnaschi, G. Degrassi, P. Slavich, A. Vicini, Higgs production via gluon fusion in the POWHEG approach in the SM and in the MSSM. JHEP **1202**, 088 (2012). arXiv:1111.2854 [hep-ph]
11. H.-L. Lai et al., New parton distributions for collider physics. Phys. Rev. D **82**, 074024 (2010). arXiv:1007.2241 [hep-ph]
12. D. de Florian, G. Ferrera, M. Grazzini, D. Tommasini, Transverse-momentum resummation: Higgs boson production at the Tevatron and the LHC. JHEP **1111**, 064 (2011). arXiv:1109.2109 [hep-ph]
13. M. Grazzini, H. Sargsyan, Heavy-quark mass effects in Higgs boson production at the LHC. JHEP **1309**, 129 (2013). arXiv:1306.4581 [hep-ph]
14. K. Hamilton, P. Nason, G. Zanderighi, MINLO: multi-scale improved NLO. JHEP **1210**, 155 (2012). arXiv:1206.3572 [hep-ph]
15. A.D. Martin, W.J. Stirling, R.S. Thorne, G. Watt, Parton distributions for the LHC. Eur. Phys. J. C **63**, 189 (2009). arXiv:0901.0002 [hep-ph]

Chapter 6
WW Measurement and Modelling

Background modelling is of paramount importance to the $H \to WW$ search. As an irreducible background, continuum $WW \to \ell\nu\ell\nu$ production dominates the 0-jet and 1-jet bins, where the majority of the sensitivity lies. Thus, even a small uncertainty in this background can have a large effect on the uncertainty in the measured signal. The process is also of general interest to electroweak phenomenology and is sensitive to anomalous triple gauge couplings (aTGCs), as seen in Fig. 6.1.

Section 6.1 describes a dedicated WW cross section measurement performed using the 2011 dataset of pp collisions at $\sqrt{s} = 7$ TeV. Then, Sect. 6.2 describes how the WW background is estimated in the phase space of the $H \to WW$ search.

6.1 Cross Section Measurement in the 0-jet Bin

This section describes the WW cross section measurement using the $\sqrt{s} = 7$ TeV dataset, published in reference [1]. As the experimental signature is the same as $H \to WW$, many aspects are shared. However, there are two key differences between the analyses:

1. The analysis described in Chap. 4 is optimised for low mass ($m_H \approx 125$ GeV) resonant WW production. It therefore requires at least one W boson to be off-shell, and consequently employs low lepton p_T thresholds ($p_T > 10$ GeV). Conversely, the measurement of non-resonant WW production is optimised for on-shell W bosons, and raises the lepton thresholds to $p_T > 20$ GeV. This reduces hadrons misidentified as leptons, and obviates the need to split the different-flavour channel into $e\mu/\mu e$.
2. The total WW cross section at $\sqrt{s} = 7$ TeV is $\sigma_{\text{NLO}} = 44.7^{+2.1}_{-1.9}$ pb, which is much larger than that of $gg \to H \to WW$ (3.3 ± 0.4 pb for $m_H = 125$ GeV). This enables backgrounds to be suppressed by tighter criteria, whilst retaining a large number of signal events. As an example, only the 0-jet bin is used.

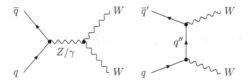

Fig. 6.1 LO Feynman diagrams for WW production, for the s-channel (*left*) and t-channel (*right*). The s-channel contains a triple gauge coupling vertex

6.1.1 Reconstruction of Physics Objects

Beam conditions in 2011 were quite different to 2012; in particular, the pile-up environment was considerably less harsh (see Fig. 3.2b). Consequently, the reconstruction of physics objects is slightly different to that described in Sect. 4.2:

Electrons

The Gaussian Sum Filter was not implemented in the 2011 reconstruction, and so the efficiency is lower (see Fig. 4.3). To reduce fakes, the cut-based *tight* identification criteria are used. The reduced pile-up allows tighter calorimeter isolation to be applied, $E_T^{\text{cone}}(0.3)/E_T < 0.14$, and this in turn allows for looser tracker isolation $p_T^{\text{cone}}(0.3)/E_T < 0.13$. Finally, the association with the primary vertex is relaxed, with the transverse impact parameter d_0 required to be within ten standard deviations of zero. The p_T threshold is raised to 20 GeV.

Muons

Differences to muon reconstruction are minimal. Slightly tighter quality criteria are applied to the ID tracks. The isolation criteria are $E_T^{\text{cone}}(0.3)/p_T < 0.14$ and $p_T^{\text{cone}}(0.3)/p_T < 0.15$. The p_T threshold is raised to 20 GeV.

Jets

A lower pile-up noise threshold is used in topo-clustering, corresponding to $\mu = 8$ (*c.f.* $\mu = 30$ in 2012). Local cluster weighting (LCW) is not performed on topo-clusters, and so jets are corrected directly from the EM scale to the Jet Energy Scale (JES). In 2011, the pile-up subtraction step of the calibration is less sophisticated, and is averaged over N_{PV} and μ rather than an event-by-event correction [2]. The jet vertex fraction criterion is also removed. A unified p_T threshold of 25 GeV is used over the entire range $|\eta| < 4.5$.

Missing transverse momentum

Calorimeter-based p_T^{inv} is used exclusively throughout the analysis.

6.1.2 Event Selection Criteria

As mentioned above, the event selection of the WW measurement can be tighter than that of the $H \rightarrow WW$ search, and does not require such stringent optimisation.

6.1 Cross Section Measurement in the 0-jet Bin

Table 6.1 Single lepton triggers employed in the 2011 WW cross section measurement

e	EF_e20_medium	14th Apr–4th Aug
	EF_e22_medium	4th Aug–22nd Aug
	EF_e22vh_medium1	7th Sep–30th Oct
μ	EF_mu18_MG	14th Apr–29th Jul
	EF_mu18_MG_medium	30th Jul–30th Oct

Trigger names are explained in Table 4.4

Table 6.2 Summary of WW event selection

$e\mu$	$ee/\mu\mu$
$p_{T,\ell}^{\text{lead}} > 25$ and $p_{T,\ell}^{\text{sublead}} > 20$	
$m_{\ell\ell} > 10$	$m_{\ell\ell} > 15$
–	$|m_{\ell\ell} - m_Z| > 15$
$E_{T,\text{rel}}^{\text{miss}} > 25$	$E_{T,\text{rel}}^{\text{miss}} > 45$
$N_{\text{jets}} = 0$	
$p_{T,\ell\ell} > 30$	

Cuts are given in GeV. The relevant observables are described in Chap. 4

Data quality

The 2011 pp dataset (see Sect. 3.2.2) is subject to data quality criteria as in 2012, though some criteria are specific to the data-taking conditions of 2011. The selected dataset corresponds to an integrated luminosity of 4.64 ± 0.18 fb^{-1}.

Trigger

The lowest unprescaled single lepton triggers were used to support $p_{T,\ell}^{\text{lead}} > 25$ GeV in the offline analysis, whilst operating on the plateau. These changed throughout the year as beam conditions changed, and are displayed in Table 6.1.

Event selection

The event selection is shown in Table 6.2 and is similar to the 0-jet $H \to WW$ selection in Table 4.5, although the topological cuts of the Higgs boson decay are obviously not applied. The major differences are the raised lepton thresholds, the raised p_T^{inv} cuts, and the lack of $\Delta\phi(\ell\ell, p_T^{\text{inv}})$ and f_{recoil} cuts.

6.1.3 Analysis Strategy

In contrast to the $H \to WW$ search, the signal region of the WW measurement has a high signal-to-background ratio. Thus, it is sufficient to simply count the number of events passing the selection, rather than fit a discriminating observable like m_T.

In order to measure the total WW cross section, it is necessary to estimate the signal acceptance of the event selection (see Sect. 6.1.4). This extrapolation from the measured phase space to the inclusive phase space introduces theoretical uncertainties. It is helpful to separate these theoretical uncertainties from the others, by measuring an intermediate cross section in a *fiducial region* of phase space, chosen to be similar to that used in the detector-level selection in order to minimise the extrapolation.

The fiducial region is defined by the criteria in Table 6.2 applied to hadron-level objects, which are now described. The MC event record is used to identify prompt leptons and neutrinos. An $E_{T,\nu\nu}$ vector is constructed from the neutrinos. Each lepton is 'dressed' by adding the four-momenta of photons within a cone of $\Delta R < 0.1$, in order to recover energy lost via QED FSR. Jets are found using particle four-momenta as inputs (cf. topo-clusters at detector-level). Muons and neutrinos are excluded from jet finding since they interact weakly with the calorimeter. Objects must pass the same p_T, η and overlap-removal criteria applied at detector-level.

The fiducial cross section is extracted from measurements using

$$\sigma_{WW}^{\text{fid}} = \frac{N_{\text{obs}} - N_{\text{bkg}}}{C_{WW} \cdot L} \quad (6.1)$$

where N_{obs} is the observed number of events passing the event selection, N_{bkg} is the expected number of background events (see Sect. 6.1.5), L is the luminosity of the dataset, and C_{WW} is the ratio of the expected number of signal events passing the detector-level selection to those passing the fiducial selection. C_{WW} accounts for detector effects such as trigger and reconstruction efficiencies, and object mismeasurement due to the finite resolution of the detector.

The total cross section can be extracted using

$$\sigma_{WW} = \frac{N_{\text{obs}} - N_{\text{bkg}}}{C_{WW} \cdot A_{WW} \cdot \text{BR} \cdot L} \quad (6.2)$$

where A_{WW} is the ratio of the expected number of signal events passing the fiducial selection to the total expected number of signal events. A_{WW} accounts for the signal acceptance of the event selection criteria. BR incorporates the branching ratios of both W bosons for the channel in question, and includes contributions from leptonic τ decays.

It is apparent from (6.1) and (6.2) that both signal and background modelling are important inputs in measuring a cross section.

6.1.4 Signal Modelling

The WW process is modelled at NLO by MC@NLO+HERWIG. The NNLO $gg \to WW$ diagrams contribute $\sim$3% to the total cross section, due to the large gluon luminosities at the LHC, and are modelled by GG2WW+HERWIG [3]. The signal acceptances C_{WW} and A_{WW} are corrected for the small mismodelling of trigger and lepton reconstruction efficiencies, through in situ tag-and-probe studies (see Sect. 4.2.3).

6.1 Cross Section Measurement in the 0-jet Bin

Table 6.3 Summary of how the experimental and theoretical uncertainties on the jet veto acceptance ϵ contribute to C_{WW}, A_{WW} and their product

Contribution to	No correction factor		Jet veto correction factor	
	Experimental	Theoretical	Experimental	Theoretical
C_{WW}	$\Delta \epsilon_{WW}^{MC}$	–	$\Delta \left(\epsilon_{WW}^{MC} / \epsilon_Z^{MC} \right)$	$\Delta \epsilon_Z^{MC}$
A_{WW}	–	$\Delta \epsilon_{WW}^{MC}$	–	$\Delta \epsilon_{WW}^{MC}$
$C_{WW} \cdot A_{WW}$	$\Delta \epsilon_{WW}^{MC}$	$\Delta \epsilon_{WW}^{MC}$	$\Delta \left(\epsilon_{WW}^{MC} / \epsilon_Z^{MC} \right)$	$\Delta \left(\epsilon_{WW}^{MC} / \epsilon_Z^{MC} \right)$

Strategies with and without the jet veto correction factor are considered

The jet veto in the event selection is responsible for the dominant uncertainties in both C_{WW} and A_{WW}. Uncertainties in the jet energy scale (JES) and jet energy resolution (JER) lead to large uncertainties in C_{WW}, since the leading jet p_T is a rapidly falling distribution. As discussed in Sect. 5.1, restricting QCD emissions via a jet veto introduces large theoretical uncertainties to A_{WW}, though these are expected to be smaller in WW than in ggF since the perturbative corrections are smaller.

For these reasons, a data-driven correction factor is included within C_{WW} in order to improve the modelling of the jet veto acceptance ϵ_{WW}. This correction factor is derived using $Z \to \ell\ell$ events. Thus, the predicted jet veto acceptance is

$$\epsilon_{WW}^{\text{pred}} = \epsilon_{WW}^{MC} \cdot \frac{\epsilon_Z^{\text{data}}}{\epsilon_Z^{MC}} . \tag{6.3}$$

Application of this correction factor effectively calibrates the MC generator to data (thus the same MC generator must be used to model ϵ_{WW}^{MC} and ϵ_Z^{MC}). As shown in Table 6.3, this enables the experimental uncertainty in the jet veto acceptance to be reduced, at the expense of introducing a theoretical uncertainty to C_{WW}. However, in the product $C_{WW} \cdot A_{WW}$ the theoretical uncertainty is also reduced.

Correction factors $\epsilon_Z^{\text{data}} / \epsilon_Z^{MC}$ are measured in control regions with high $Z \to \ell\ell$ purity. Events are selected with $p_{T,\ell}^{\text{lead}} > 25$ GeV, $p_{T,\ell}^{\text{sublead}} > 20$ GeV and $|m_{\ell\ell} - m_Z| < 15$ GeV in the ee and $\mu\mu$ channels. The correction factor for the $e\mu$ channel is taken as the average of the other two. The correction factors are 0.957, 0.954 and 0.956 for the ee, $\mu\mu$ and $e\mu$ channels respectively. Statistical uncertainties are negligible compared to the experimental and theoretical uncertainties.

Uncertainties in ϵ_{WW}^{MC} and ϵ_Z^{MC} due to the JES are evaluated by increasing and decreasing jet energies by one standard deviation, and using the average of the absolute deviations. The JES uncertainty is determined during the in situ calibration [4]. Similarly, uncertainties due to the JER are evaluated by increasing the JER by one standard deviation. The JER uncertainty is determined using other in situ techniques [5]. These uncertainties are displayed in Table 6.4.

Theoretical uncertainties are evaluated at hadron-level by changing some aspect of the MC modelling and measuring the effect upon the jet veto acceptance ϵ. These uncertainties are displayed in Table 6.4, and their estimation is described below.

Table 6.4 Relative uncertainties in the WW and Z jet veto acceptances, and in their ratio

	Sources of relative uncertainty					Total	
	JES	JER	Scale	PDF	PS/UE	Exper.	Theor.
$e\mu$ channel							
$\epsilon_{WW}^{MC} = 0.662$	4.6%	2.4%	5.3%	1.6%	0.5%	5.1%	5.6%
$\epsilon_{Z}^{MC} = 0.793$	3.8%	2.6%	2.4%	0.8%	0.4%	4.6%	2.6%
$\epsilon_{WW}^{MC}/\epsilon_{Z}^{MC} = 0.835$	0.7%	0.2%	3.4%	0.9%	0.1%	0.7%	3.5%
ee channel							
$\epsilon_{WW}^{MC} = 0.652$	4.8%	2.0%	5.3%	1.6%	0.5%	5.2%	5.6%
$\epsilon_{Z}^{MC} = 0.798$	3.7%	2.4%	2.4%	0.8%	0.4%	4.4%	2.6%
$\epsilon_{WW}^{MC}/\epsilon_{Z}^{MC} = 0.817$	1.1%	0.4%	3.4%	0.9%	0.1%	1.2%	3.5%
$\mu\mu$ channel							
$\epsilon_{WW}^{MC} = 0.655$	4.8%	2.1%	5.3%	1.6%	0.5%	5.2%	5.6%
$\epsilon_{Z}^{MC} = 0.788$	4.0%	2.8%	2.4%	0.8%	0.4%	4.9%	2.6%
$\epsilon_{WW}^{MC}/\epsilon_{Z}^{MC} = 0.831$	0.8%	0.7%	3.4%	0.9%	0.1%	1.0%	3.5%

Results for the ee, $\mu\mu$ and $e\mu$ channels are shown separately, with ϵ_Z^{MC} for the $e\mu$ channel taken as the average of the other two

Uncertainties due to higher order corrections are evaluated using the combined inclusive method described in Sect. 5.1.2. The renormalisation and factorisation scales are independently varied in the range $\mu_0/2 \leq \mu_R, \mu_F \leq 2\mu_0$, where μ_0 is the default scale for the process in question,[1] whilst observing the constraint $1/2 \leq \mu_R/\mu_F \leq 2$. The largest deviation is used as the uncertainty.

Uncertainties due to parton distribution functions (PDFs) are evaluated in two ways, and added in quadrature. The jet veto acceptance predicted with the CT10 PDF is compared to that predicted with the MSTW2008 PDF [6]. The set of PDF eigenvectors corresponding to 90 % CL of the CT10 fit are also used.

Uncertainties due to the parton shower (PS), hadronisation and underlying event (UE) models are evaluated by recomputing the jet veto acceptances with POWHEGBOX events, and comparing the result when showered by HERWIG and by PYTHIA6.

By substituting uncertainties from Table 6.4 into the two strategies outlined in Table 6.3, it is clear that the jet veto correction factor offers a significant reduction in uncertainty. Considering the σ_{WW} extraction from the $e\mu$ channel, the contribution

---

[1] The default scales used by MC@NLO are determined by $\mu_0^2 = m_{\ell\ell}^2 + p_{T,\ell\ell}^2$ for Z production and $\mu_0^2 = (m_{\ell\nu}^2 + p_{T,\ell\nu}^2 + m_{\ell'\nu'}^2 + p_{T,\ell'\nu'}^2)/2$ for WW production.

6.1 Cross Section Measurement in the 0-jet Bin

Table 6.5 The signal acceptances C_{WW}, A_{WW} and $C_{WW} \cdot A_{WW}$

	$e\mu$	ee	$\mu\mu$	All
C_{WW} ($\times 100$)	50.5 ± 1.6	40.3 ± 1.7	68.7 ± 2.1	52.0 ± 1.7
Trigger effciency	0.3%	0.1%	0.6%	0.4%
Lepton effciency	1.4%	2.9%	0.7%	1.3%
Lepton p_T scale and resolution	0.6%	0.9%	0.8%	0.5%
Jet energy scale and resolution				
Jet veto	0.7%	1.2%	1.0%	1.0%
Other cuts	0.5%	0.6%	0.5%	0.5%
p_T^{inv} modelling	0.4%	0.5%	0.2%	0.2%
PDFs, μ_R and μ_F scales				
Acceptance	0.3%	0.7%	0.7%	0.3%
Jet veto correction factor	2.6%	2.6%	2.6%	2.6%
A_{WW} ($\times 100$)	15.9 ± 0.9	7.5 ± 0.4	8.1 ± 0.5	11.9 ± 0.7
PDFs, μ_R and μ_F scales				
Jet veto	5.6%	5.6%	5.6%	5.6%
Other cuts	1.1%	1.0%	1.0%	1.0%
$C_{WW} \cdot A_{WW}$ ($\times 100$)	8.03 ± 0.33	3.02 ± 0.15	5.56 ± 0.22	6.19 ± 0.25
PDFs, μ_R and μ_F scales				
Jet veto	3.5%	3.5%	3.5%	3.5%

A breakdown of the relative uncertainties from different sources is also shown. The contribution to $\Delta(C_{WW} \cdot A_{WW})$ of the theoretical uncertainty in the jet veto is given, in order to explicitly exhibit the cancellations between ΔC_{WW} and ΔA_{WW}

of the experimental uncertainty in ϵ_{WW} reduces from 5.1 to 0.7 %, while the theoretical uncertainty reduces from 5.6 to 3.5 %. However, it does introduce a theoretical uncertainty to σ_{WW}^{fid} of 2.6 %.

The signal acceptances A_{WW}, C_{WW} and $C_{WW} \cdot A_{WW}$ are displayed in Table 6.5. The $\mu\mu$ channel has the best reconstruction efficiency and the event selection of the $e\mu$ channel has the highest yield. The uncertainties arising from different systematic sources are also shown, with those associated with the jet veto separated for clarity. The JES and JER uncertainties in the jet veto acceptance are approximated to be uncorrelated with those of the other cuts. The same is true of the theoretical uncertainties in A_{WW}.

6.1.5 Background Modelling

The background estimation techniques of the $H \to WW$ search are more sophisticated than those of the WW measurement at $\sqrt{s} = 7$ TeV. For this reason, the techniques used in the WW measurement are only described briefly here, whereas those of the $H \to WW$ search shall be discussed in detail in Sect. 6.2 and Chap. 7.

Table 6.6 MC generators used to model backgrounds to the WW measurement

Process	MC generator
$t\bar{t}$	MC@NLO+HERWIG
tW, tb, tbq	ACERMC+PYTHIA6
$W+$jet, Z/γ^*, $W\gamma$	ALPGEN+HERWIG
$W\gamma^*$	MADGRAPH+PYTHIA6
WZ, ZZ	HERWIG

Although many backgrounds are modelled by a data-driven technique, there is often some underlying dependence upon MC. For this reason, Table 6.6 states the MC generators used to model each process.

W+jet and dijet

The $W+$jet background comprises events where a jet is misidentified as a lepton. The rate of misidentification, or "fake rate", is poorly modelled by MC, and so a data-driven *fake factor method* is used. The dijet background, where two jets fake leptons, is very small and is simultaneously estimated by this method.

A $W+$jet anti-ID region (AR) is defined similarly to the signal region (SR), but where one of the leptons is replaced with an anti-identified lepton, ℓ. The ℓ objects are ensured to be highly contaminated with jets by loosening the selection criteria and vetoing leptons passing the full identification. Anti-ID electrons, e, have looser calorimeter isolation criteria and do not have to pass the *tight* identification criteria. Anti-ID muons, μ, have looser calorimeter isolation criteria and the tracker isolation removed completely. The predicted $W+$jet background is the measured number of events in the AR, scaled by a fake factor f_ℓ:

$$N_{W+\text{jet}}^{\text{pred,SR},ee} = f_e \cdot N_{e\ell}^{\text{data,AR}} \quad (6.4)$$

$$N_{W+\text{jet}}^{\text{pred,SR},\mu\mu} = f_\mu \cdot N_{\mu\mu}^{\text{data,AR}} \quad (6.5)$$

$$N_{W+\text{jet}}^{\text{pred,SR},e\mu} = f_e \cdot N_{\mu\ell}^{\text{data,AR}} + f_\mu \cdot N_{\mu e}^{\text{data,AR}}. \quad (6.6)$$

The fake factor f_ℓ is defined as the ratio of efficiencies for jets passing the lepton ID criteria to jets passing the anti-ID criteria. Each fake factor is measured from dijet events as a function of p_T and η:

$$f_\ell = \frac{N_\ell^{\text{data}}}{N_{\ell}^{\text{data}}}. \quad (6.7)$$

To avoid bias, the dijet events were selected with a very loose (but prescaled) trigger, without identification or isolation criteria applied. A Z veto ($|m_{\ell\ell} - m_Z| > 15$ GeV) and a W veto ($m_T > 30$ GeV) reject most of the prompt leptons, and residual electroweak contamination is subtracted using MC.

The dominant source of uncertainty stems from the fact that the fake factor is derived from dijet events but applied to $W+$jet events. This process dependence

6.1 Cross Section Measurement in the 0-jet Bin

is estimated with MC. It would be preferable to measure the fake factor from Z+jet events, though this was statistically limited in the $\sqrt{s} = 7$ TeV dataset. This point shall be revisited in Sect. 7.1 when considering the $W + $ jet background to the $H \to WW$ search.

Top

The top background is suppressed by the jet veto, and its yield is estimated by a data-driven *template method*. This constrains the top N_{jets} distribution within a b-tagged control region (CR), subtracting contamination via a data-driven template fit. Then the 0-jet yield is extrapolated from the CR to the SR using MC.

In detail, an extended signal region (ESR) is defined by removing the jet veto and the $p_{T,\ell\ell}$ cut. The top CR is a subset of the ESR, requiring at least one b-tagged jet with $p_T > 20$ GeV. The top N_{jets} distribution, $\mathcal{T}$, in the ESR is predicted by measuring $\mathcal{T}$ in the CR and extrapolating using MC:

$$\mathcal{T}_{\text{top}}^{\text{pred,ESR}} = \frac{\mathcal{T}_{\text{top}}^{\text{MC,ESR}}}{\mathcal{T}_{\text{top}}^{\text{MC,CR}}} \left(\mathcal{T}^{\text{data,CR}} - K_{\text{non-top}}^{\text{fit}} \cdot \mathcal{T}_{\text{non-top}}^{\text{MC,CR}} \right). \quad (6.8)$$

The non-top contamination in the CR is described by MC and scaled by a data-driven normalisation factor $K_{\text{non-top}}^{\text{fit}}$. This $K_{\text{non-top}}^{\text{fit}}$ is assumed to be the same in the CR and the ESR, and is constrained by an N_{jets} fit in the ESR, i.e.

$$\mathcal{T}^{\text{data,ESR}} = \mathcal{T}_{\text{top}}^{\text{pred,ESR}} + K_{\text{non-top}}^{\text{fit}} \cdot \mathcal{T}_{\text{non-top}}^{\text{MC,ESR}}. \quad (6.9)$$

Note that this fit depends upon $K_{\text{non-top}}^{\text{fit}}$ twice: directly and via $\mathcal{T}_{\text{top}}^{\text{pred,ESR}}$. The fit yields $K_{\text{non-top}}^{\text{fit}} = 1.07 \pm 0.03$. Finally, the 0-jet bin of $\mathcal{T}_{\text{top}}^{\text{pred,ESR}}$ is extrapolated to the SR using MC.

The small number of 0-jet events in the top CR leads to a large statistical uncertainty in this background. The systematic uncertainty is dominated by uncertainties in the (mis)tag efficiency of the b-tagging algorithm.

Z/γ^*

The Z/γ^* background is suppressed by the Z mass veto and the $E_{T,\text{rel}}^{\text{miss}}$ and $p_{T,\ell\ell}$ cuts, though remains a significant background to the ee and $\mu\mu$ channels. Since MC might mismodel the $E_{T,\text{rel}}^{\text{miss}}$ distribution in Z/γ^* events, this background is normalised to data in a control region (CR) that includes the $E_{T,\text{rel}}^{\text{miss}}$ cut.

The Z/γ^* CR is defined similarly to the signal region (SR), with the $p_{T,\ell\ell}$ cut inverted. The number of events in the CR is measured, the non-Z/γ^* contamination is subtracted via MC, and the result is extrapolated to the SR using MC:

$$N_{Z/\gamma^*}^{\text{pred,SR}} = \frac{N_{Z/\gamma^*}^{\text{MC,SR}}}{N_{Z/\gamma^*}^{\text{MC,CR}}} \left(N^{\text{data,CR}} - N_{\text{non-}Z/\gamma^*}^{\text{MC,CR}} \right). \quad (6.10)$$

The dominant uncertainties are due to the small number of events in the CR and the energy scale of the soft terms in the $E_{\mathrm{T,rel}}^{\mathrm{miss}}$ (see Sect. 4.2.7).

Non-WW diboson

The non-WW diboson backgrounds (WZ, $W\gamma^*$, ZZ, $W\gamma$) are estimated purely from MC simulation (see Table 6.6). The cross section of each process is calculated at NLO with MCFM [7]. The dominant uncertainties are theoretical uncertainties in the cross section and JES uncertainties in the jet veto.

6.1.6 Experimental Results

The observed and expected $p_{\mathrm{T},\ell}^{\mathrm{lead}}$ and $\Delta\phi(\ell,\ell)$ distributions in the signal region are displayed in Fig. 6.2, and the observed number of events passing the event selection is shown for each channel in Table 6.7. The expected number of events is also shown, with the signal and background contributions estimated as described in Sects. 6.1.4 and 6.1.5, respectively. Systematic uncertainties in the Z/γ^* and non-WW diboson backgrounds are considered correlated (since they are both dominated by uncertainties in the calorimeter energy scale), and those in the other backgrounds are considered uncorrelated. A comparison of data to expectation suggests that the measured cross section shall be higher than predicted.

The fiducial and total cross sections are calculated for each channel with (6.1) and (6.2) respectively, using the values in Tables 6.5 and 6.7 and an integrated luminosity of $L = 4.64 \pm 0.18$ fb^{-1}. Systematic uncertainties in the signal acceptance and the total background estimation are assumed uncorrelated, and propagate to the cross sections via

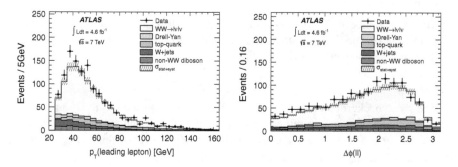

Fig. 6.2 The $p_{\mathrm{T},\ell}^{\mathrm{lead}}$ (*left*) and $\Delta\phi(\ell,\ell)$ (*right*) distributions for the selected WW candidates, including $e\mu$, ee and $\mu\mu$ channel events

6.1 Cross Section Measurement in the 0-jet Bin

Table 6.7 The number of events observed and expected in the 4.6 fb^{-1} dataset in each signal region

	$e\mu$	ee	$\mu\mu$	All
Observed	821	174	330	1325
Expected	744 ± 24 ± 57	169 ± 12 ± 16	280 ± 16 ± 20	1192 ± 31 ± 87
WW	538 ± 3 ± 45	100 ± 2 ± 9	186 ± 2 ± 15	824 ± 4 ± 69
Background	206 ± 24 ± 35	68 ± 12 ± 13	94 ± 15 ± 13	369 ± 31 ± 53
Top	87 ± 23 ± 13	22 ± 12 ± 3	32 ± 14 ± 5	141 ± 30 ± 22
W+jet	70 ± 2 ± 31	21 ± 1 ± 11	7 ± 1 ± 3	98 ± 2 ± 43
Z/γ *	5 ± 2 ± 1	12 ± 3 ± 3	34 ± 6 ± 10	51 ± 7 ± 12
Diboson	44 ± 2 ± 6	13 ± 1 ± 2	21 ± 1 ± 2	78 ± 2 ± 10

A breakdown of the expected signal and background contributions is also shown, with statistical and systematic uncertainties. The WW signal is normalised to the NLO cross section of 44.7 pb

$$\left(\frac{\Delta\sigma_{WW}^{\text{fid}}}{\sigma_{WW}^{\text{fid}}}\right)_{\text{syst}}^2 = \left(\frac{\Delta C_{WW}}{C_{WW}}\right)^2 + \left(\frac{\Delta N_{\text{bkg}}}{N_{\text{obs}} - N_{\text{bkg}}}\right)^2 + \left(\frac{\Delta L}{L}\right)^2 \quad (6.11)$$

$$\left(\frac{\Delta\sigma_{WW}}{\sigma_{WW}}\right)_{\text{syst}}^2 = \left(\frac{\Delta(C_{WW} \cdot A_{WW})}{C_{WW} \cdot A_{WW}}\right)^2 + \left(\frac{\Delta N_{\text{bkg}}}{N_{\text{obs}} - N_{\text{bkg}}}\right)^2 + \left(\frac{\Delta L}{L}\right)^2. \quad (6.12)$$

The relative statistical uncertainty in each channel is simply $\sqrt{N_{\text{obs}}}/(N_{\text{obs}} - N_{\text{bkg}})$.

The combined total cross section from the three channels is obtained by maximising the likelihood function of a Poisson process

$$\mathcal{L}\left(\sigma_{WW}|N_{\text{obs}}^{e\mu}, N_{\text{obs}}^{ee}, N_{\text{obs}}^{\mu\mu}\right) = \text{Pois}\left(N_{\text{obs}}^{e\mu}; \sigma_{WW}\right) \cdot \text{Pois}\left(N_{\text{obs}}^{ee}; \sigma_{WW}\right) \cdot \text{Pois}\left(N_{\text{obs}}^{\mu\mu}; \sigma_{WW}\right)$$

$$= \prod_{i=1}^{3} \frac{e^{-\left(N_{\text{sig}}^i + N_{\text{bkg}}^i\right)} \cdot \left(N_{\text{sig}}^i + N_{\text{bkg}}^i\right)^{N_{\text{obs}}^i}}{N_{\text{obs}}^i!} \quad (6.13)$$

where $N_{\text{sig}}^i = \sigma_{WW} \cdot C_{WW}^i \cdot A_{WW}^i \cdot \text{BR}^i \cdot L$ and the product over i corresponds to the three channels. The best fit $\hat{\sigma}_{WW}$ is that which maximises the likelihood, and the statistical uncertainty is obtained by finding where $\ln \mathcal{L}(\hat{\sigma}_{WW} \pm \Delta\sigma_{WW}) = \ln \mathcal{L}(\hat{\sigma}_{WW}) - \frac{1}{2}$.

The systematic uncertainty in the combined cross section is calculated with (6.12), again using the values in Tables 6.5 and 6.7. The top and non-WW diboson systematic uncertainties are considered correlated between all three channels (since the b-tagging efficiency, JES and total diboson cross sections are independent of lepton flavour). The W + jet and Z/γ^* systematic uncertainties are treated as uncorrelated between the ee and $\mu\mu$ channels, though the $ee + \mu\mu$ combination is considered correlated with the $e\mu$ channel (since the f_e and f_μ fake factors are determined

Table 6.8 Measured fiducial and total WW cross sections extracted from each signal region

	Fiducial cross section (fb)		Total cross section (pb)	
	Measured	Predicted	Measured	Predicted
$e\mu$	$262.3 \pm 12.3 \pm 20.7 \pm 10.2$	231.4 ± 15.7	$51.1 \pm 2.4 \pm 4.2 \pm 2.0$	44.7 ± 2.0
ee	$56.4 \pm 6.8 \pm 9.8 \pm 2.2$	54.6 ± 3.7	$46.9 \pm 5.7 \pm 8.2 \pm 1.8$	44.7 ± 2.0
$\mu\mu$	$73.9 \pm 5.9 \pm 6.9 \pm 2.9$	58.9 ± 4.0	$56.7 \pm 4.5 \pm 5.5 \pm 2.2$	44.7 ± 2.0
All			$51.9 \pm 2.0 \pm 3.9 \pm 2.0$	44.7 ± 2.0

Theoretical predictions are shown for comparison. The uncertainties in measured quantities are statistical, systematic and luminosity, respectively

independently, and the $Z/\gamma^* \rightarrow \ell\ell$ uncertainty is related to lepton energy mismeasurement).

The measured cross sections are shown in Table 6.8. The combined total cross section is 51.9 ± 2.0 (stat) ± 3.9 (syst) ± 2.0 (lumi) pb, which is slightly higher than the theoretical prediction of 44.7 ± 2.0 pb. However, the discrepancy is not significant. If contributions from $H \rightarrow WW$ and VBF WW production are included as backgrounds, the measured cross section reduces to 50.5 ± 2.0 (stat) ± 3.9 (syst) ± 2.0 (lumi) pb.

6.2 Background Estimation for $H \rightarrow WW$ Search

Non-resonant WW production is an irreducible background to the $H \rightarrow WW$ search, and is the dominant background contribution following the event selection described in Chap. 4. Thus, it is critical that it is accurately estimated.

The WW process is modelled at NLO by POWHEGBOX+PYTHIA6. It is necessary to use POWHEGBOX since MC@NLO does not feature "singly resonant diagrams", where the mass of the dilepton + dineutrino system is that of a single Z boson. This occurs in $Z/\gamma^* \rightarrow \ell\ell$ events where a lepton radiates a W boson, and the final state is indistinguishable from the WW process. Since the $H \rightarrow WW$ search is sensitive to off-shell W bosons, it is important to include these diagrams. The POWHEGBOX+PYTHIA6 event generator is used as it is found to better describe the experimental data in many exclusive observables, when compared to POWHEGBOX+PYTHIA8. This is likely to be related to the POWHEGBOX+PYTHIA8 matching issues mentioned in Sect. 2.2.6. Also, the ATLAS underlying event AUET2B tune [8] was found to be overtuned to dijet data and consequently its description of electroweak processes suffered. For this reason, the updated Perugia 2011C PYTHIA6 tune [9] was used. NNLO $gg \rightarrow WW$ diagrams are modelled by GG2WW+HERWIG [3].

To estimate the WW contribution to the signal region (SR), a data-driven technique is employed whereby MC is used to extrapolate from a high-purity control region (CR) to the SR:

6.2 Background Estimation for $H \to WW$ Search

$$N_{WW}^{\text{pred,SR}} = \alpha_{WW} \cdot \left(N^{\text{data,CR}} - N_{\text{non-}WW}^{\text{pred,CR}}\right) \quad (6.14)$$

$$\alpha_{WW} = N_{WW}^{\text{MC,SR}} / N_{WW}^{\text{MC,CR}} \quad (6.15)$$

where $N_{\text{non-}WW}^{\text{pred,CR}}$ is determined by dedicated methods (see Chap. 7). The CR is chosen to reduce $N_{\text{non-}WW}^{\text{pred,CR}}$ to avoid bias, whilst maintaining a large $N^{\text{data,CR}}$ to minimise the statistical uncertainty in $N_{WW}^{\text{pred,SR}}$. The MC-based extrapolation α_{WW} introduces experimental and theoretical uncertainties to $N_{WW}^{\text{pred,SR}}$.

As explained in Sect. 4.3.4, the resonant $H \to WW$ and non-resonant WW processes are distinguished by the scalar nature of the Higgs boson, which affects the decay topology of the W bosons. This causes $H \to WW$ events to have a small dilepton opening angle, and consequently low $m_{\ell\ell}$ and $\Delta\phi(\ell, \ell)$. It is therefore intuitive to define the WW CR at high $m_{\ell\ell}$, with a relaxed $\Delta\phi(\ell, \ell)$ requirement.

Since jet binning can introduce large uncertainties, a WW CR is defined for each of the 0-jet and 1-jet bins separately (see Table 6.9). Unfortunately, it is not possible to define a WW CR in the $\geq$2-jet bin with sufficient purity owing to the large top background. Thus, the WW background estimation in the $\geq$2-jet bin is MC-based (see Sect. 6.2.3).

Since the $ee/\mu\mu$ channels are dominated by $Z/\gamma^* \to \ell\ell$, they are extrapolated from $e\mu/\mu e$ CRs. The selection criteria follow those of the SRs (see Table 4.5), with different $m_{\ell\ell}$ and $\Delta\phi(\ell, \ell)$ cuts. However, the $p_{\text{T},\ell}^{\text{sublead}}$ threshold is raised from 10 to 15 GeV in order to reduce contamination from the W + jet background. A validation region (VR) is also defined in the 0-jet bin with $m_{\ell\ell} > 110\,\text{GeV}$, to test the extrapolation from the CR.

Figure 6.3 exhibits the excellent description of the shapes of distributions in the CRs, following application of the normalisation factors.

Table 6.9 Event selection criteria of the WW control regions (not used in the $\geq$2-jet bin)

$e\mu/\mu e$	
$p_{\text{T},\ell}^{\text{lead}} > 22$ and $p_{\text{T},\ell}^{\text{sublead}} > 15$	
$p_{\text{T}}^{\text{miss,corr}} > 20$	
0-jet bin	1-jet bin
$p_{\text{T},\ell\ell} > 30$	$|m_{\tau\tau} - m_Z| > 25$
$\Delta\phi(\ell\ell, p_{\text{T}}^{\text{inv}}) > \pi/2$	$\max(m_{\text{T},\ell}) > 50$
–	$N_{b\text{-jets}} = 0$
$55 < m_{\ell\ell} < 110$	$m_{\ell\ell} > 80$
$\Delta\phi(\ell, \ell) < 2.6$	–

Cuts on energy, momentum and mass are given in GeV, and angular cuts are given in radians. The relevant observables are described in Chap. 4

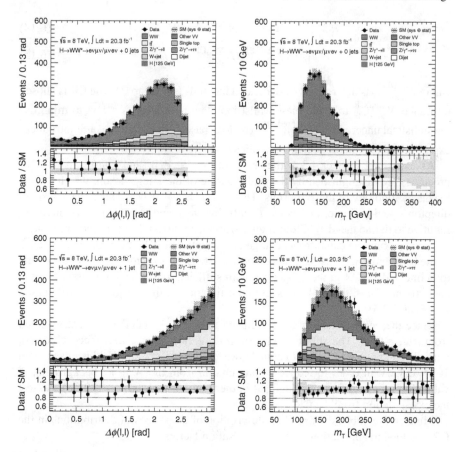

Fig. 6.3 The $\Delta\phi(\ell,\ell)$ (*left*) and m_T (*right*) distributions in the WW control regions of the 0-jet (*top*) and 1-jet (*bottom*) bins. Normalisation factors are applied

6.2.1 Theoretical Uncertainties in α_{WW}

Theoretical uncertainties in the extrapolation parameters α_{WW} are evaluated at hadron-level (i.e. before detector simulation) by changing some aspect of the MC modelling and measuring the effect upon each α_{WW}. This is done using NLO WW MC. Uncertainties due to $gg \to WW$ diagrams are $\sim 7\%$, but have small impact on the uncertainty in the measured signal since these diagrams only contribute $\sim 5\%$ ($\sim 7\%$) of the WW background in the 0-jet (1-jet) bin SRs.

The hadron-level event selection criteria used to evaluate these uncertainties are similar to the detector-level criteria of the SRs (see Table 4.5) and CRs (see Table 6.9), except the b-jet veto and f_{recoil} cuts are omitted. Hadron-level object definitions are the same as those used for the WW measurement (see Sect. 6.1.3), with updated p_T requirements.

6.2 Background Estimation for $H \to WW$ Search

Table 6.10 Theoretical uncertainties in the WW extrapolation parameter α_{WW} for each signal region used in the fitting procedure, and also for the validation region

	$m_{\ell\ell}$ (GeV)	$p_{T,\ell}^{sublead}$ (GeV)	Scale (%)		PDF (%)		PS/UE (%)	NLO-PS (%)
			QCD	EW	Collab.	68 % CL		
$ee/\mu\mu$ channels								
0-jet	12–55	>10	0.8	+0.1	0.5	1.0	−1.2	+2.4
1-jet	12–55	>10	0.8	−2.1	0.5	0.7	−2.3	+3.8
$e\mu/\mu e$ channels								
0-jet	10–30	10–15	0.7	+1.2	0.9	0.2	+2.2	+0.4
		15–20	1.2	+0.7	0.8	0.2	+1.7	+0.9
		>20	0.7	−0.3	0.5	0.3	−1.9	+3.1
	30–55	10–15	0.7	+0.8	0.8	0.1	+1.5	+0.5
		15–20	0.8	+0.5	0.7	0.2	+1.0	+1.0
		>20	0.8	−0.4	0.4	0.5	−2.4	+3.9
1-jet	10–30	10–15	3.1	−0.9	0.5	0.1	−2.4	−3.4
		15–20	1.6	−1.5	0.5	0.1	−3.0	+0.7
		>20	1.0	−2.8	0.6	0.2	−3.6	+5.3
	30–55	10–15	3.2	−0.9	0.5	0.1	−2.0	+1.9
		15–20	1.5	−1.6	0.4	0.1	−3.0	+2.4
		>20	1.3	−2.7	0.5	0.4	−3.1	+5.6
Validation region ($e\mu/\mu e$ channels)								
0-jet	>110	>15	0.6	+1.6	0.6	2.0	+4.3	−5.1

Four sources of theoretical uncertainty are considered:

- higher order corrections (QCD and EW),
- PDFs,
- parton shower, hadronisation and underlying event models,
- NLO-PS matching scheme.

Uncertainties due to QCD higher order corrections are evaluated via independent variation of renormalisation and factorisation scales in the range $\mu_0/2 \leq \mu_R, \mu_F \leq 2\mu_0$, where $\mu_0 = m_{\ell\nu\ell'\nu'}$, whilst observing the constraint $1/2 \leq \mu_R/\mu_F \leq 2$. These are evaluated with AMC@NLO and validated with MCFM. The largest deviation is used as the uncertainty.

Uncertainties due to EW higher order corrections are evaluated by reweighting POWHEGBOX events (based upon the kinematics of the initial state and diboson system) to include NLO EW corrections. Such corrections are derived in reference [10].

Uncertainties due to PDFs are evaluated in two ways, which are added in quadrature. Predictions with the MSTW 2008 [6] and NNPDF 2.3 [11] PDF sets are compared to those with the CT10 [12] PDF sets, and the maximum deviation is found. Also, the set of PDF eigenvectors corresponding to 90 % CL of the CT10 fit were used

Fig. 6.4 WW m_T shape systematic uncertainties in each 0-jet signal region for the $e\mu/\mu e$ channels. The fit shown is the sum in quadrature of the three individual fits, and successfully envelopes the uncertainty sources

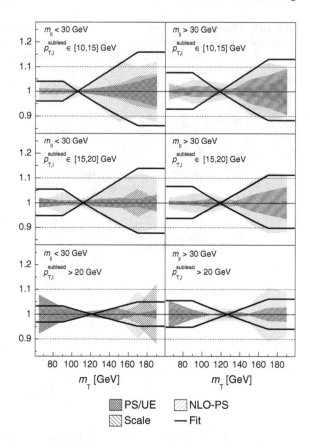

to evaluate an uncertainty, which was then rescaled to 68 % CL. These are evaluated using AMC@NLO.

Uncertainties due to the PS, hadronisation and UE models are evaluated by comparing POWHEGBOX showered by PYTHIA6 (nominal) and by HERWIG. Uncertainties due to the NLO-PS matching scheme are evaluated by comparing POWHEG-BOX+HERWIG to AMC@NLO+HERWIG.

The α_{WW} theoretical uncertainties for each SR are shown in Table 6.10 (experimental uncertainties are discussed in Chap. 8). Although small, they have a large effect in the final fit because the WW background yield is much larger than the signal yield in the SR. Extrapolation uncertainties for the VR are also shown. The observed number of events in the WW VR is 10 % higher than expected, corresponding to a 0.65σ deviation when statistical and systematic uncertainties are considered. This agreement supports the extrapolation uncertainties assigned to the SRs.

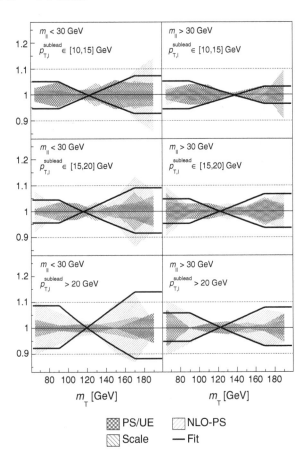

Fig. 6.5 WW m_T shape systematic uncertainties in each 1-jet signal region for the $e\mu/\mu e$ channels. The fit shown is the sum in quadrature of the three individual fits, and successfully envelopes the uncertainty sources

6.2.2 m_T Shape Modelling

Theoretical uncertainties in the shape of the m_T distribution are also investigated, as they affect the fit. Uncertainties due to scale, PS/UE and NLO-PS choices are considered within each signal region separately, using the methods described above.

Each uncertainty is parametrised by fitting the ratio of the m_T shapes, and then symmetrising the fit to produce "up" and "down" variations. The m_T distributions are normalised to unit integral in order to probe only shape uncertainties. A linear fit is used in the central m_T region, and a constant is used in the low-m_T and high-m_T tails of the distribution where statistical fluctuations dominate.

These fits allow the hadron-level m_T distribution of the WW background to be reweighted to the "up" and "down" variations. In this way, the m_T shape uncertainty is treated as a nuisance parameter in the $H \to WW$ fitting procedure (see Chap. 8). The uncertainties for the split signal regions of the $e\mu/\mu e$ channels are displayed in Figs. 6.4 and 6.5 for the 0-jet and 1-jet bins, respectively.

6.2.3 WW Background in the ≥2-jet Bin

POWHEGBOX relies upon the parton shower to produce the second emission; consequently events with two or more hard jets are likely to be poorly modelled. For this reason, SHERPA is used to describe the WW background in the ≥2-jet bin, including up to three partons in the matrix element using LO ME-PS merging (see Sect. 2.2.3).

Two kinds of diagrams can be identified for the $WW + 2$ jets process: EW production (zero QCD vertices at LO) and QCD production (two QCD vertices at LO), where VBF WW production is included in the former. The two production types are simulated separately by SHERPA, and this causes their interference to be neglected. Interference with VBF Higgs boson production is also neglected. The effect of each interference is investigated and treated as a systematic uncertainty.

Uncertainties due to the choice of μ_R and μ_F scale are considered, and also a modelling uncertainty is evaluated by comparing the SHERPA yield to that of MADGRAPH+PYTHIA6. The uncertainty in QCD $WW + 2$ jets production is ~20%, and the uncertainty in EW $WW + 2$ jets production is ~10%.

References

1. ATLAS Collaboration, Measurement of W^+W^- production in pp collisions at $\sqrt{s}$=7 TeV with the ATLAS detector and limits on anomalous WWZ and $WW\gamma$ couplings. Phys. Rev. D **87**, 112001 (2013). arXiv:1210.2979 [hep-ex]
2. ATLAS Collaboration, Pile-up corrections for jets from proton-proton collisions at $\sqrt{s} = 7$ TeV in ATLAS in 2011, ATLAS-CONF-2012-064 (2012)
3. T. Binoth, M. Ciccolini, N. Kauer, M. Kramer, Gluon-induced W-boson pair production at the LHC. JHEP **0612**, 046 (2006). arXiv:hep-ph/0611170
4. ATLAS Collaboration, Jet energy scale and its systematic uncertainty in proton-proton collisions at $\sqrt{s} = 7$ TeV with ATLAS 2011 data, ATLAS-CONF-2013-004 (2013)
5. ATLAS Collaboration, Jet energy resolution in proton-proton collisions at $\sqrt{s} = 7$ TeV recorded in 2010 with the ATLAS detector. Eur. Phys. J. C **73**, 2306 (2013). arXiv:1210.6210 [hep-ex]
6. A.D. Martin, W.J. Stirling, R.S. Thorne, G. Watt, Parton distributions for the LHC. Eur. Phys. J. C **63**, 189 (2009). arXiv:0901.0002 [hep-ph]
7. J.M. Campbell, R.K. Ellis, C. Williams, Vector boson pair production at the LHC. JHEP **1107**, 018 (2011). arXiv:1105.0020 [hep-ph]
8. ATLAS Collaboration, Further ATLAS tunes of PYTHIA6 and Pythia 8, ATL-PHYS-PUB-2011-014 (2011)
9. P.Z. Skands, Tuning Monte Carlo generators: the Perugia tunes. Phys. Rev. D **82**, 074018 (2010). arXiv:1005.3457 [hep-ph]
10. A. Bierweiler, T. Kasprzik, J.H. Kühn, S. Uccirati, Electroweak corrections to W-boson pair production at the LHC. JHEP **1211**, 093 (2012). arXiv:1208.3147 [hep-ph]
11. R.D. Ball et al., Parton distributions with LHC data. Nucl. Phys. B **867**, 244 (2013). arXiv:1207.1303 [hep-ph]
12. H.-L. Lai et al., New parton distributions for collider physics. Phys. Rev. D **82**, 074024 (2010). arXiv:1007.2241 [hep-ph]

Chapter 7
Other Backgrounds

In addition to the *WW* background, several other processes contribute significant background. As they are often difficult to model accurately in the $H \to WW$ phase space, they are estimated by sophisticated data-driven methods. Even so, there is usually some underlying dependence upon MC, which is summarised in Table 7.1.

This chapter describes the estimation of non-*WW* backgrounds: $W+$jet and dijet in Sect. 7.1, non-*WW* diboson in Sect. 7.2, top in Sect. 7.3, and Z/γ^* in Sect. 7.4.

7.1 $W +$ jet and Dijet

$W +$ jet events contribute to the background when a jet is misidentified as a lepton, and dijet events contribute when two jets are misidentified as leptons. This is due to leptonic decays of heavy flavour hadrons, hadronic showers mimicking the electromagnetic showers, or punchthrough into the muon spectrometer. Although the fake rates are very low, these backgrounds are significant because the $W +$ jet and dijet cross sections are many orders of magnitude larger than those of Higgs boson production. Since these fake rates are sensitive to effects that are difficult to accurately model (such as jet flavour composition, jet substructure and hadronic shower shapes), a data-driven *fake factor method* is used to estimate these backgrounds.

As these two backgrounds have large uncertainties, their suppression is critically important. The electron and muon selection criteria are chosen to be tighter at low p_T (see Sect. 4.2), in order to reduce the fake rates where these backgrounds are largest. The dijet background is additionally rejected by requiring significant p_T^{inv} and by the $\max(m_{T,\ell}) > 50\,\text{GeV}$ cut in the 1-jet bin of the $e\mu/\mu e$ channels.

7.1.1 The Fake Factor Method

The fake factor method defines two new objects: anti-identified electrons $\tilde{e}$ and muons $\tilde{\mu}$, collectively known as anti-identified leptons $\tilde{\ell}$. Loosened selection criteria ensure

Table 7.1 MC generators used to model backgrounds to the $H \to WW$ search

Process	MC generator ($\geq$2-jet bin)
WW	POWHEGBOX+PYTHIA 6, GG2WW+HERWIG (SHERPA)
Top	POWHEGBOX+PYTHIA 6, ACERMC+PYTHIA 6
W + jet, Z/γ^*, $W\gamma$	ALPGEN+HERWIG
WZ	POWHEGBOX+PYTHIA 8
ZZ	POWHEGBOX+PYTHIA 8, GG2ZZ+HERWIG
$W\gamma^*$, $Z\gamma^*$, $Z\gamma$	SHERPA

These are used with the data-driven techniques described in the text

they are highly contaminated by jets, while a veto on identified leptons ℓ removes overlap between ℓ and $\not\ell$ objects (see Sect. 7.1.2). The fake factor f_ℓ is then defined as the ratio of efficiencies (i.e. fake rates) for jets passing the ℓ criteria to passing the $\not\ell$ criteria.

In the following discussion, a *sample* shall refer to a collection of events featuring a specific set of objects, e.g. the signal sample $\mathcal{N}_{\ell\ell,\mathrm{OS}}$ refers to $\ell\ell$ events with opposite-sign charges (the $H \to WW$ analysis uses three signal samples: $\ell\ell = ee+\mu\mu$, $e\mu$ and μe). A *region* shall refer to events whose objects satisfy a set of criteria (also known as cuts), e.g. the signal region refers to events passing the criteria in Table 4.5.

A dilepton sample $\mathcal{N}_{\ell\ell}$, a W + jet control sample $\mathcal{N}_{\ell\not\ell}$ and a dijet control sample $\mathcal{N}_{\not\ell\not\ell}$ are defined, each with opposite-sign (OS) and same-sign (SS) charge variants. Note that $\mathcal{N}$ indicates a sample to which cuts may be applied, and an $\not\ell$ is treated as an ℓ in such cuts when appropriate. Each sample has contributions from W + jet and dijet events, and events with prompt leptons from the hard scatter or photon conversions (labelled EW):

$$\mathcal{N}_{\ell\ell,i} = \mathcal{N}_{\ell\ell,i}^{W+\mathrm{jet}} + \mathcal{N}_{\ell\ell,i}^{\mathrm{dijet}} + \mathcal{N}_{\ell\ell,i}^{\mathrm{EW}} \tag{7.1}$$

$$\mathcal{N}_{\ell\not\ell,i} = \mathcal{N}_{\ell\not\ell,i}^{W+\mathrm{jet}} + \mathcal{N}_{\ell\not\ell,i}^{\mathrm{dijet}} + \mathcal{N}_{\ell\not\ell,i}^{\mathrm{EW}} \tag{7.2}$$

$$\mathcal{N}_{\not\ell\not\ell,i} = \mathcal{N}_{\not\ell\not\ell,i}^{W+\mathrm{jet}} + \mathcal{N}_{\not\ell\not\ell,i}^{\mathrm{dijet}} + \mathcal{N}_{\not\ell\not\ell,i}^{\mathrm{EW}} \tag{7.3}$$

where $i = $ OS, SS. $\mathcal{N}_{\ell\ell,i}$ is dominated by $\mathcal{N}_{\ell\ell,i}^{\mathrm{EW}}$, $\mathcal{N}_{\ell\not\ell,i}$ is dominated by $\mathcal{N}_{\ell\not\ell,i}^{W+\mathrm{jet}}$ and $\mathcal{N}_{\not\ell\not\ell,i}$ is dominated by $\mathcal{N}_{\not\ell\not\ell,i}^{\mathrm{dijet}}$.

The purpose of the fake factor method is to estimate the $\mathcal{N}_{\ell\ell,\mathrm{OS}}^{W+\mathrm{jet}}$ and $\mathcal{N}_{\ell\ell,\mathrm{OS}}^{\mathrm{dijet}}$ contributions to the $\mathcal{N}_{\ell\ell,\mathrm{OS}}$ signal sample. The SS versions are required in the non-WW diboson background estimation (see Sect. 7.2). It does this in a data-driven way by using the W + jet and dijet control samples, subtracting the expected contaminations, and then multiplying by a data-driven fake factor f_ℓ (defined earlier):

7.1 W + jet and Dijet

$$N_{\ell\ell,i}^{\text{pred,dijet}} = \left(N_{\ell\!\!\!/\ell\!\!\!/,i}^{\text{data}} - N_{\ell\!\!\!/\ell\!\!\!/,i}^{\text{MC},W+\text{jet}} - N_{\ell\!\!\!/\ell\!\!\!/,i}^{\text{MC,EW}}\right) \cdot f_{\ell|\ell\!\!\!/}^{\text{pred,dijet}} \cdot f_{\ell|\ell\!\!\!/}^{\text{pred,dijet}} \quad (7.4)$$

$$N_{\ell\ell\!\!\!/,i}^{\text{pred,dijet}} = \left(N_{\ell\!\!\!/\ell\!\!\!/,i}^{\text{data}} - N_{\ell\!\!\!/\ell\!\!\!/,i}^{\text{MC},W+\text{jet}} - N_{\ell\!\!\!/\ell\!\!\!/,i}^{\text{MC,EW}}\right) \cdot \sum_{\ell} f_{\ell|\ell\!\!\!/}^{\text{pred,dijet}} \quad (7.5)$$

$$N_{\ell\ell,i}^{\text{pred},W+\text{jet}} = \left(N_{\ell\ell\!\!\!/,i}^{\text{data}} - N_{\ell\ell\!\!\!/,i}^{\text{pred,dijet}} - N_{\ell\ell\!\!\!/,i}^{\text{MC,EW}}\right) \cdot f_{\ell,i}^{\text{pred},W+\text{jet}} \quad (7.6)$$

where $i = $ OS, SS. In the $e\mu/\mu e$ channels, there are in fact two terms because an electron or a muon could be the fake. The dijet contamination to the $W+$jet control sample $N_{\ell\ell\!\!\!/,i}^{\text{pred,dijet}}$ is data-driven from the dijet control sample. As discussed later, the fake factors are determined by the flavour composition of the jets, and thus depend upon the process. Also, $f_\ell^{\text{pred,dijet}}$ depends upon whether the other object in the event is an ID lepton ($f_{\ell|\ell}^{\text{pred,dijet}}$) or an anti-ID lepton ($f_{\ell|\ell\!\!\!/}^{\text{pred,dijet}}$). This shall be discussed in Sect. 7.1.6. Finally, note that $f_\ell^{\text{pred},W+\text{jet}}$ depends upon whether the event is OS or SS. This shall be discussed in Sect. 7.1.5.

An advantage of this sample-based fake factor method compared to that used in the WW cross section measurement (see Sect. 6.1.5) is that it provides a background estimation in regions other than the signal region.

The rest of this section on the $W+$jet and dijet backgrounds shall be spent describing the anti-identification lepton selection criteria (Sect. 7.1.2), the measurement of fake factors in experimental data (Sects. 7.1.3 and 7.1.4), and MC-based corrections to these fake factors to improve the background estimations (Sects. 7.1.5 and 7.1.6).

7.1.2 Lepton Anti-Identification Criteria

The $\ell\!\!\!/$ selection criteria are loosened with respect to the ℓ selection criteria in order to accept more jets. An explicit veto upon ℓ objects avoids overlap between samples.

Relative to the e definition in Sect. 4.2.3, anti-ID electrons of any p_T must fail the *medium* identification criteria (though instead must have $n_{\text{pixel}}^{\text{hit}} + n_{\text{SCT}}^{\text{hit}} \geq 4$). Also, the tracker and calorimeter isolation are loosened to $p_T^{\text{cone}}(0.3)/E_T < 0.16$ and $E_T^{\text{cone}}(0.3)/E_T < 0.30$.

Relative to the μ definition in Sect. 4.2.4, anti-ID muons have their transverse impact parameter d_0 requirement removed. Also, the tracker isolation is removed and the calorimeter isolation is loosened to $E_T^{\text{cone}}(0.3)/p_T < 0.15$ for $p_T \in [10, 15]$ GeV, $E_T^{\text{cone}}(0.3)/p_T < 0.25$ for $p_T \in [15, 20]$ GeV and $E_T^{\text{cone}}(0.3)/p_T < 0.30$ for $p_T > 20$ GeV.

7.1.3 Dijet Fake Factor Measurement

In situ fake factor measurements are made using dijet events. This involves counting the numbers of ℓ and $\ell\!\!\!/$ objects in a dijet control region (CR), subtracting the expected contamination of prompt leptons from W and Z boson events, and calculating the ratio $f_\ell = N_\ell / N_{\not\ell}$. $f_\ell^{\text{data,dijet}}$ is measured as a function of p_T and η.

Following data quality requirements, events are selected using very loose (no isolation or electron identification criteria), but highly prescaled, lepton triggers. To reduce the effects of prescaling, different triggers were used for different p_T ranges, and a trigger containing electron identification criteria was added to aid measurement of N_ℓ. In the f_e measurement, the EF_e5_etcut (0.012 pb^{-1}) and EF_e5_medium1 (0.24 pb^{-1}) triggers were used for $p_T < 20$ GeV and the EF_g24_etcut (2.1 pb^{-1}) trigger was used for $p_T > 24$ GeV. In the f_μ measurement, the EF_mu6 (0.94 pb^{-1}) trigger was used for $p_T < 15$ GeV and the EF_mu15 (23 pb^{-1}) trigger was used for $p_T > 15$ GeV. The trigger naming scheme is explained in the caption of Table 4.4.

The dijet CR requires events to have a jet with $p_T > 15$ GeV (see Sect. 4.2.5 for jet selection) balancing the triggered lepton object, $\Delta\phi(\ell, j) > 0.7$. To suppress contamination from W boson events we require $m_{T,\ell} < 30$ GeV, and to suppress the Z boson background we veto events with a lepton pair satisfying $\left|m_{\ell\ell} - m_Z\right| < 13$ GeV. Note that these criteria apply to both ℓ and $\not\ell$ objects. Normalisation factors for the MC predictions of the residual W and Z boson backgrounds are derived by inverting the corresponding veto.

The measured electron and muon fake factors are shown in Fig. 7.1. The uncertainty is dominated by uncertainties in the subtracted electroweak contamination (which is conservatively scaled up and down by 20 %). $f_\ell^{\text{data,dijet}}$ is used in the dijet background estimation, as described in Sect. 7.1.6.

7.1.4 $Z + jet$ Fake Factor Measurement

In situ fake factor measurements are also made using Z + jet events. This involves counting the numbers of ℓ and $\not\ell$ objects in a Z + jet control region (CR), subtracting the expected prompt leptons and photon conversions from electroweak contamination ($Z\gamma$, ZZ, $Z\gamma^*$, $W\gamma$, WZ, $W\gamma^*$), and calculating their ratio $f_\ell = N_\ell / N_{\not\ell}$.

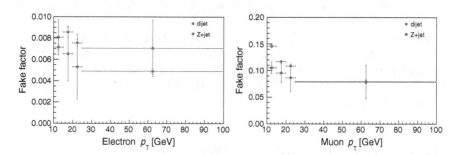

Fig. 7.1 The fake factor measured in dijet (*red*) and Z + jet (*blue*) events versus p_T, for electrons (*left*) and muons (*right*) [1]. The *error bars* include statistical uncertainties and uncertainties in the background subtraction

Following data quality requirements, events are selected using unprescaled lepton triggers. This is possible because the triggered object is a lepton from the Z boson decay, and the p_T threshold can therefore be relatively high. In the f_e measurement, the `EF_e24vhi_medium1` and `EF_e60_medium1` triggers support $p_T >$ 25 GeV. In the f_μ measurement, the `EF_mu24i_tight` and `EF_mu36_tight` triggers are used with the dilepton `EF_mu18_tight_mu8_EFFS` trigger to support $p_T >$ 22 GeV. The trigger naming scheme is explained in the caption of Table 4.4.

The Z + jet CR requires a pair of same-flavour and oppositely charged ℓ objects to reconstruct the Z boson mass, $81 < m_{\ell\ell} < 107$ GeV. These leptons are excluded from the f_ℓ calculation. Electroweak contamination is suppressed by cuts on additional ℓ and $\not\ell$ objects: ZZ is rejected by a veto on $76 < m_{\ell\ell} < 107$ GeV for additional dilepton systems, and WZ is rejected by $m_{T,\ell} < 30$ GeV. Residual diboson backgrounds ($Z\gamma$, ZZ, $Z\gamma^*$, $W\gamma$, WZ, $W\gamma^*$) are subtracted using MC predictions.

The measured electron and muon fake factors are shown in Fig. 7.1. The uncertainty is dominated by statistical uncertainty. For this reason, $f_\ell^{\text{data},Z+\text{jet}}$ is measured as a function of p_T only, and the η dependence is injected from $f_\ell^{\text{data,dijet}}$. The uncertainty due to electroweak subtraction is also significant (estimated by varying the diboson cross sections), because the contamination to the Z + jet CR is not negligible. Although $f_\ell^{\text{data},Z+\text{jet}}$ has larger uncertainties than $f_\ell^{\text{data,dijet}}$, it shall be used in the W + jet background estimation. This is because jets in Z + jet and W + jet events are expected to have similar flavour composition, and therefore similar fake factors (see Sect. 7.1.5).

7.1.5 W + jet Background Estimation

The W + jet background to the opposite-sign (OS) and same-sign (SS) dilepton samples are estimated from the W + jet control sample, using (7.6). The W + jet control sample $\mathcal{N}_{\ell\not\ell}$ contains events with one ID lepton and one anti-ID lepton, selected from events passing the data quality criteria and triggers specified in Sect. 4.3. Generally, it is the lepton from the W boson decay that fires the single lepton triggers. Contamination is small, and is estimated by the fake factor method for dijet events (see Sect. 7.1.6) and MC for other processes.

The fake factor of a process is determined by the jet flavour composition of that process. Specifically, f_e is larger for heavy flavour because the electron particle identification is tuned to reject light flavour jets, and f_μ is larger for light flavour because the d_0 criterion suppresses long-lived heavy flavour hadrons.

Since we use both OS and SS samples, it is useful to consider the jet flavour composition of the W + jet process in each sample. There are diagrams like $qg \to Wq'$, where the outgoing quark has OS charge to the W boson (e.g. $W + c$),[1] and

---

[1] It should be emphasised that $qg \to W + b$ is highly suppressed by the near unity of V_{tb} in the CKM matrix and the negligible top contribution to the incoming PDFs.

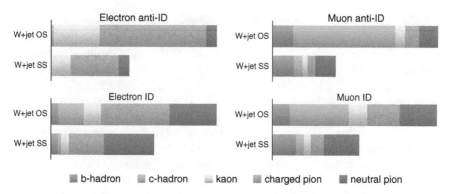

Fig. 7.2 The jet flavour composition of fake leptons, for opposite-sign (OS) and same-sign (SS) $W + $ jet events. Each SS axis is scaled to the cross section of the corresponding OS axis

there are diagrams like $q\bar{q}' \rightarrow Wg$, where the gluon subsequently splits to a $q\bar{q}$ pair (e.g. $W + b\bar{b}$). Therefore, in terms of the W and quark charges, the former diagrams contribute to OS only and the latter contribute to OS and SS equally. However, the samples are categorised according to the relative signs of the leptons, and so the extent to which the quark charge is preserved in the reconstructed lepton is important. When a hadron decays leptonically, its sign is preserved in the lepton. Without such a non-prompt lepton, the sign of the jet charge is more likely to be inverted in the reconstructed lepton. In the case of photon conversions from $\pi^0 \rightarrow \gamma\gamma$, there is no asymmetry. Thus, considering the above points, there is a strong OS/SS asymmetry in c-jets, a mild asymmetry in light flavour jets, and no asymmetry in b-jets and $\pi^0 \rightarrow \gamma\gamma$ (see Fig. 7.2).

MC-based corrections are applied to the measured $Z + $ jet fake factors in order to account for the different jet flavour compositions of the OS and SS $W + $ jet processes

$$f_{\ell,i}^{\text{pred},W+\text{jet}}(p_T, \eta) = f_\ell^{\text{data},Z+\text{jet}}(p_T) \cdot \frac{f_\ell^{\text{data,dijet}}(p_T, \eta)}{f_\ell^{\text{data,dijet}}(p_T)} \cdot \frac{f_{\ell,i}^{\text{MC},W+\text{jet}}}{f_\ell^{\text{MC},Z+\text{jet}}} \quad (7.7)$$

where $i = $ OS, SS. The injection of η-dependence from the dijet fake factor is also shown in (7.7). The $Z + $ jet fake factor is used because $Z + $ jet has a similar jet flavour composition to $W + $ jet. The correction factors are derived with ALPGEN+PYTHIA 6, and compared to those of ALPGEN+HERWIG and POWHEGBOX+PYTHIA 8 to obtain a systematic uncertainty. The correction factors are 0.99 ± 0.05 (stat) ± 0.19 (syst) for OS electrons, 1.00 ± 0.08 (stat) ± 0.21 (syst) for OS muons, 1.25 ± 0.08 (stat) ± 0.30 (syst) for SS electrons and 1.40 ± 0.14 (stat) ± 0.47 (syst) for SS muons.

The $W + $ jet background is dominated by uncertainties in the fake factor. These are split into components that are correlated and uncorrelated between $f_{\ell,\text{OS}}^{W+\text{jet}}$ and $f_{\ell,\text{SS}}^{W+\text{jet}}$. Components correlated between OS and SS largely cancel in the same-sign control region method of estimating the non-WW diboson background (see Sect. 7.2.1). This control region also offers validation of the $W + $ jet background estimation.

7.1.6 Dijet Background Estimation

The dijet backgrounds to the dilepton and $W+$ jet samples are estimated from the dijet control sample, using (7.4) and (7.5) respectively. The dijet control sample $\mathcal{N}_{\ell\ell}$ contains events with two anti-ID leptons, selected from events passing the data quality criteria and triggers specified in Sect. 4.3. They are accepted by the dilepton triggers since these have looser lepton selections. Contamination from $W+$ jet and other processes is estimated with MC, though is generally small.

The dijet fake factor $f_\ell^{\text{data,dijet}}$ is measured in events containing a balancing jet (passing the offline jet selection). However, in (7.4) and (7.5) the fake factors are applied to events containing another anti-ID lepton ($f_{\ell|\ell}^{\text{pred,dijet}}$), or as if there is an ID lepton in the event ($f_{\ell|\ell}^{\text{pred,dijet}}$). This can heavily bias the jet flavour composition of the selected events, and this in turn can affect the fake factor of interest. For example, if the other object is a muon, this increases the probability that the event contains heavy flavour jets. Consequently, f_e would increase and f_μ would decrease (see Sect. 7.1.5).

MC-based corrections are applied to the measured dijet fake factor in order to account for the correlation between f_ℓ and the other object in the event

$$f_{\ell|\ell}^{\text{pred,dijet}}(p_T, \eta) = f_\ell^{\text{data,dijet}}(p_T, \eta) \cdot \frac{f_{\ell|\ell}^{\text{MC,dijet}}}{f_{\ell|j}^{\text{MC,dijet}}} \qquad (7.8)$$

$$f_{\ell|\ell}^{\text{pred,dijet}}(p_T, \eta) = f_\ell^{\text{data,dijet}}(p_T, \eta) \cdot \frac{f_{\ell|\ell}^{\text{MC,dijet}}}{f_{\ell|j}^{\text{MC,dijet}}} . \qquad (7.9)$$

The dijet background has a very small contribution to the $H \to WW$ signal regions, but has large uncertainties dominated by the MC-based corrections. It is difficult to define a high purity dijet validation region in the dilepton sample, though good agreement with experimental data is observed in the low p_T^{inv} regions where this background is enhanced (see Fig. 7.3).

7.2 Non-WW Diboson

The non-WW diboson background comprises the $W\gamma$, $W\gamma^*$, WZ and ZZ processes (in order of contribution to the signal region). $W\gamma$ events feature a prompt photon that passes the electron selection, via an asymmetric conversion (e^+e^- production). The other processes have signatures of $\ell\nu\ell\ell$, $\ell\ell\ell\ell$ or $\ell\ell\nu\nu$, and usually contribute when one or more leptons fail object selection (e.g. $p_T < 10\,\text{GeV}$ or $|\eta| > 2.5$).

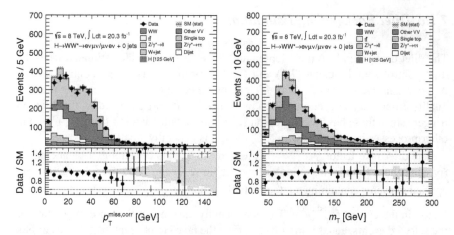

Fig. 7.3 The $p_T^{miss,corr}$ (*left*) and m_T (*right*) distributions in the $e\mu/\mu e$ channels of same-sign dilepton events. Selected events have passed the $p_{T,\ell}^{lead} > 22\,\text{GeV}$, $m_{\ell\ell} > 10\,\text{GeV}$, $N_{jets} = 0$ and $\Delta\phi\left(\ell\ell, p_T^{inv}\right) > \pi/2$ criteria

7.2.1 Same-Sign Control Region

In non-WW diboson backgrounds, a symmetry is expected to exist between opposite-sign (OS) and same-sign (SS) dilepton events; this is particularly true in the $e\mu/\mu e$ channels, where asymmetric final states are reduced. Conversely, the WW, top and Z/γ^* backgrounds only contribute to the OS sample. Finally, the W+jet background displays a partial OS/SS symmetry, as described in Sect. 7.1.5. Thus, SS events can be used to estimate the non-WW diboson background in the OS signal region, whilst validating the W + jet estimation.

An SS control region (CR) is defined using identical criteria to the OS signal region (SR). In the language of Sect. 7.1, the SS CR events are in the SR of the $\mathcal{N}_{\ell\ell,SS}$ sample. This is used to determine the normalisation of the non-WW diboson background, whilst the shapes of observables used in the fitting procedure (i.e. m_T, $m_{\ell\ell}$ and $p_{T,\ell}^{sublead}$) are modelled by MC. This is equivalent to

$$N_{VV}^{\text{pred,SR}} = \alpha_{VV} \cdot \left(N^{\text{data,CR}} - N_{\text{non-}VV}^{\text{pred,CR}}\right) \quad (7.10)$$

$$\alpha_{VV} = N_{VV}^{\text{MC,SR}} / N_{VV}^{\text{MC,CR}} \quad (7.11)$$

where $VV = W\gamma + W\gamma^* + WZ + ZZ$, and $N_{\text{non-}VV}^{\text{pred,CR}}$ is dominated by W + jet. The extrapolation α_{VV} represents the MC predictions for shapes of observables, since the OS/SS symmetry implies that the total normalisation is largely unchanged.

The SS CR method is only used in the 0-jet and 1-jet bins of the $e\mu/\mu e$ channels, using a combined $e\mu + \mu e$ channel to define the CR. All other signal regions estimate the non-WW diboson background using MC only.

7.2 Non-WW Diboson

Uncertainties in the normalisation of the constituent processes will cancel if the composition of the non-WW diboson background is the same in SS and OS events; this is modelled by MC. However, uncertainties in the shapes of distributions remain, since these are not constrained by the SS CR method.[2] Additionally, the uncertainty component of the OS $W+$ jet background that is correlated between OS and SS will cancel in the SS CR method; an increase in the SS $W+$ jet background will be compensated by a decrease in the non-WW diboson background prediction, and vice versa.

Figure 7.4 validates the MC shape modelling of the fit observables in the SS CRs, following application of the normalisation factors. Good agreement with experimental data is observed. Note that these SS distributions are not directly used, since only the total number of events in the SS CRs are used in (7.10).

7.2.2 $W\gamma$

$W\gamma$ events enter the dilepton sample when the photon fakes an electron. This is usually caused by an asymmetric $\gamma \to e^+e^-$ conversion, where only one electron is reconstructed in the tracker and calorimeter. This background is suppressed by the electron identification criteria, which require a hit in the first pixel layer and no conversion vertex (see Sect. 4.2.3).

$W\gamma$ is modelled by ALPGEN+HERWIG and normalised to the NLO cross section calculated with MCFM. Modelling is tested with a SS dilepton sample of $e\mu/\mu e$ events, but where the electron object is required to have a conversion vertex and no hit in the first pixel layer (i.e. the photon conversion rejection criteria are inverted). Validation regions (VRs) are defined in the 0-jet and 1-jet bins, using the corresponding signal region cuts. Experimental data are well described, as seen in Fig. 7.5.

As the conversion and pixel hit criteria are inverted in the VR, their SR modelling is not tested in Fig. 7.5. Unfortunately, it is not possible to define a high-purity $W\gamma$ VR when including these criteria. Instead, a $Z \to \mu\mu\gamma$ VR is used to test this modelling. Events are selected with an OS muon pair with $p_{T,\mu}^{\text{lead}} > 22\,\text{GeV}$ and $p_{T,\mu}^{\text{sublead}} > 10\,\text{GeV}$, and an additional electron candidate with $p_{T,e} > 10\,\text{GeV}$. Low mass resonances are vetoed by $m_{\mu\mu} > 12\,\text{GeV}$, and events featuring QED FSR are selected by $\left|m_{\mu\mu e} - m_Z\right| < 15\,\text{GeV}$. The selected events are ~55% $Z\gamma$ and ~45% $WZ/Z\gamma^*$. A mismodelling is found depending upon $p_{T,e}$ and so a conversion systematic uncertainty is derived: 25% for 10–15 GeV, 18% for 15–20 GeV and 5% for >20 GeV.

---

[2] Shape uncertainties are also introduced by normalisation uncertainties in the individual processes comprising the non-WW diboson background, which lead to a composition uncertainty.

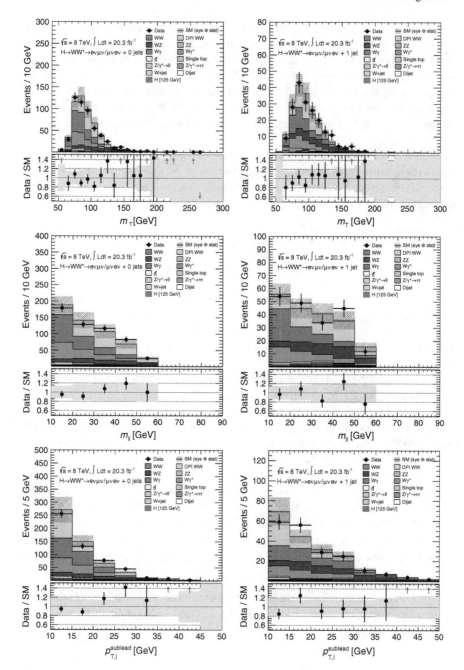

Fig. 7.4 The m_T (*top*), $m_{\ell\ell}$ (*middle*) and $p_{T,\ell}^{\text{sublead}}$ (*bottom*) distributions in the 0-jet (*left*) and 1-jet (*right*) same-sign control regions. Normalisation factors are applied

7.2 Non-WW Diboson

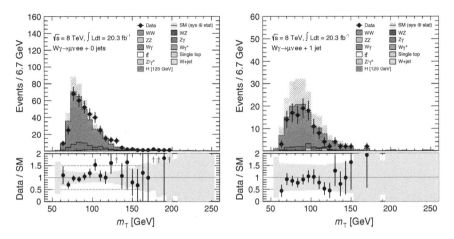

Fig. 7.5 The m_T distribution in the 0-jet (*left*) and 1-jet (*right*) $W\gamma$ validation regions [1]. The *shaded error band* includes statistical and theoretical uncertainties, in addition to those associated with conversion modelling

7.2.3 WZ and $W\gamma^*$

Both the WZ and $W\gamma^*$ processes result in a $\ell\nu\ell'\ell'$ final state, peaking in $m_{\ell'\ell'}$ at m_Z and $2m_{\ell'}$ respectively (due to the different Z and γ propagators). Their interference at intermediate $m_{\ell'\ell'}$ is non-negligible, and so these processes are generated together.

It is technically difficult to generate MC events with very low $m_{\ell'\ell'}$. For this reason, the phase space is generated in two parts: high mass "WZ" events with $m_{\ell'\ell'} > 7\,\text{GeV}$ and low mass "$W\gamma^*$" events with $m_{\ell'\ell'} < 7\,\text{GeV}$.[3] In ambiguous cases ($\ell = \ell'$), the lowest mass opposite-sign same-flavour dilepton pair is used to split the phase space. WZ is modelled at NLO by POWHEGBOX+PYTHIA 8.

$W\gamma^*$ is technically difficult to model because it involves integrating over a phase space with extremely low mass; the threshold for production is $m_{\ell'\ell'} = 2m_{\ell'}$, which is 1.022 MeV for $\ell\nu ee$. In previous analysis iterations [2], $W\gamma^*$ was modelled by MADGRAPH+PYTHIA 6. However, this implementation is not designed to produce gauge-invariant results when $m_{\ell'\ell'} \ll \Gamma_W$ [3], and numerical limitations result in wildly unphysical events in this region of phase space (e.g. leptons with $p_T \sim 1\,\text{TeV}$). In an attempt to resolve this problem, events with $m_{\ell'\ell'} < 3\,\text{MeV}$ were removed from the MC sample, and events with $3\,\text{MeV} < m_{\ell'\ell'} < 10\,\text{MeV}$ were reweighted in order to recover the cross section. Even with this fix, this background was associated with a modelling uncertainty of 40% (evaluated by comparing to SHERPA).

SHERPA can produce gauge-invariant results for the full mass range, since it employs a complex mass scheme [4]. A simple LO SHERPA sample underestimates the FSR phase space with $p_{T,\ell'\ell'} > m_W/2$. For this reason, $W\gamma^*$ is modelled by

---

[3] It is these "WZ" and "$W\gamma^*$" MC samples that are referred to as the WZ and $W\gamma^*$ backgrounds, though each sample contains both the WZ and $W\gamma^*$ processes and all interference.

ME-PS merging with up to one additional parton (see Sect. 2.2.3). It is not currently possible to include further partons in the ME-PS merging over the full mass range.

The SHERPA sample is normalised using an NLO K-factor of 0.94 ± 0.07 (scale), calculated with MCFM. Due to technical limitations, this is calculated in a high mass region $m_{\ell'\ell'} \in [0.5, 7]\,\text{GeV}$ and then extrapolated down in mass. It is calculated with the $p_{\text{T},\ell}^{\text{lead}} > 22\,\text{GeV}$ and $p_{\text{T},\ell}^{\text{sublead}} > 10\,\text{GeV}$ criteria.

Since the $W\gamma^*$ acceptance of the lepton p_T criteria is strongly related to N_jets and because the $H \to WW$ analysis itself is jet binned, the jet multiplicity distribution must be well modelled. For this reason, the N_jets distribution is reweighted to that of a SHERPA ME-PS merged sample with up to two additional partons, using corrections of 0.91 ± 0.06 (scale) in the 0-jet bin, 1.09 ± 0.33 (scale) in the 1-jet bin, and 2.0 ± 0.5 (scale) in the $\geq$2-jet bin. Again, due to technical limitations, this is calculated in a high mass region $m_{\ell'\ell'} \in [0.5, 7]\,\text{GeV}$ and then extrapolated down in mass. It is also calculated with the $p_{\text{T},\ell}^{\text{lead}} > 22\,\text{GeV}$ and $p_{\text{T},\ell}^{\text{sublead}} > 10\,\text{GeV}$ criteria.

Two sources of theoretical uncertainties in the signal region acceptances are considered: higher order corrections, estimated by varying μ_R and μ_F as described elsewhere in this thesis; and modelling uncertainties, evaluated by comparing the SHERPA ME-PS samples with $\leq$1 parton (the default) and $\leq$2 partons in a high mass region $m_{\ell'\ell'} \in [0.5, 7]\,\text{GeV}$. The uncertainty in the selection acceptance are negligible compared to the theoretical uncertainties in the K-factor and jet-bin correction. Scale uncertainties in the shape of the m_T distribution are also evaluated.

The $W\gamma^*$ modelling is tested in a $W\gamma^* \to e\nu\mu\mu$ validation region (VR). Since $\Delta R(\mu, \mu)$ is generally small, the muon isolation criteria are altered: tracks associated with other muons are removed from the p_T^cone definition, and the calorimeter isolation is loosened to $E_\text{T}^\text{cone}(0.3)/p_\text{T} < 0.4$ for $p_\text{T} < 15\,\text{GeV}$ (cf. Sect. 4.2.4). Events are selected with $p_{\text{T},e} > 22\,\text{GeV}$, $p_{\text{T},\mu}^\text{lead} > 10\,\text{GeV}$ and $p_{\text{T},\mu}^\text{sublead} > 3\,\text{GeV}$. Then, the VR is defined by $m_{\mu\mu} < 7\,\text{GeV}$, $\left|m_{\mu\mu} - m_{J/\psi}\right| > 100\,\text{MeV}$, $E_\text{T,rel}^\text{miss} > 20\,\text{GeV}$ and $\max(\Delta\phi(\ell, \ell)) < 2.8$. The experimental data in the VR is well described by SHERPA, as seen in Fig. 7.6, and supports the normalisation used. It appears that N_jets is underestimated, though this effect is within theoretical uncertainties (which are not shown in Fig. 7.6).

It should be noted that the $W\gamma^*$ VR does not test the modelling of $W\gamma^* \to \ell\nu ee$, which can contribute background events via an additional mechanism: both electrons can be reconstructed as a single electron when $\Delta R(e, e)$ is very small. However, the good agreement in the same-sign control region suggests that this is well-modelled.

7.2.4 ZZ and Zγ^*

Similar to Sect. 7.2.3, the ZZ and Zγ^* processes share a $\ell\ell\ell'\ell'$ final state, and are modelled together in order to include their interference. Again, the phase space is split: high mass "ZZ" events with $m_{\ell\ell} > 4\,\text{GeV}$ and $m_{\ell'\ell'} > 4\,\text{GeV}$, and low mass

7.2 Non-WW Diboson

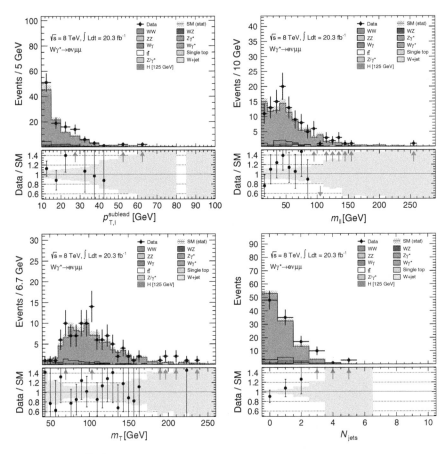

Fig. 7.6 The $p_{T,\ell}^{sublead}$ (*top left*), $m_{\ell\ell}$ (*top right*), m_T (*bottom left*) and N_{jets} (*bottom right*) distributions in the $W\gamma^*$ validation region [1]. Leptonic observables are defined with the electron and the leading muon

"$Z\gamma^*$" events with $m_{\ell\ell} > 4\,\text{GeV}$ and $m_{\ell'\ell'} < 4\,\text{GeV}$.[4] The corresponding "$\gamma^*\gamma^*$" process is not modelled due to technical limitations. In ambiguous cases ($\ell = \ell'$), the lowest mass opposite-sign same-flavour dilepton pair is used to split the phase space. The $\ell\ell\nu'\nu'$ final state is also considered, though its contribution is negligible.

ZZ is modelled at NLO by POWHEGBOX+PYTHIA 8 and the NNLO $gg \to ZZ$ diagrams are also modelled by GG2ZZ+HERWIG. $Z\gamma^*$ is modelled by SHERPA and is normalised using an NLO K-factor of 0.88 ± 0.05 (scale), calculated with MCFM. Neither background contributes significantly to the signal region. However, it is important to model the $Z\gamma^*$ background in the $Z\gamma$ validation region (see Sect. 7.2.2) and when measuring the Z + jet fake factor (see Sect. 7.1.4).

---

[4] It is these "ZZ" and "$Z\gamma^*$" MC samples that are referred to as the ZZ and $Z\gamma^*$ backgrounds, though each sample contains the ZZ, $Z\gamma^*$ and $\gamma^*\gamma^*$ processes and all interference.

7.3 Top

The leading and subleading contributions to the top background are the $t\bar{t}$ and tW processes, respectively. These are irreducible backgrounds, in that they both exhibit the opposite-sign dilepton + p_T^{inv} experimental signature. This signature results from the top decays, BR $(t \to Wb) \approx 100\,\%$ (occurring before hadronisation), followed by leptonic W boson decays.

The b-jets present in top events motivate the jet binning of the $H \to WW$ analysis (see Fig. 4.12). Jets with a p_T threshold of 25 (30) GeV in the central (forward) region are used for this binning (see Sect. 4.2.5). The top background is further discriminated by counting the number of b-tagged jets with $p_T > 20\,\text{GeV}$ in the central region, using an algorithm with a tagging efficiency of 85 % (see Sect. 4.2.6). In the 1-jet and $\geq$2-jet bins, the top background is suppressed by vetoing events with b-tagged jets.

$t\bar{t}$, tW and s-channel single top (tb) are modelled by POWHEGBOX+PYTHIA 6, while t-channel single top (tbq) is modelled by ACERMC+PYTHIA 6. However, the jet binning and b-tagged jet veto introduce large modelling uncertainties, and so data-driven techniques are used to estimate this background.

7.3.1 0-jet Bin Estimation

In the 0-jet bin, the top background is very small because both b-jets must fail the jet selection. It is estimated by the data-driven *jet veto survival probability method*.

An extended signal region (ESR) is defined by the pre-selection in Sect. 4.3.3, with an additional $\Delta\phi\,(\ell, \ell) < 2.8$ criterion to suppress the $Z/\gamma^* \to \tau\tau$ background. The aim is to estimate the number of events in the ESR passing the jet veto $N_{\text{top}}^{\text{pred,ESR,0j}}$, and then extrapolate to the 0-jet signal region (SR) with MC

$$N_{\text{top}}^{\text{pred,SR,0j}} = \alpha_{\text{top}}^{0j} \cdot \epsilon_{0,\text{top}}^{\text{pred,ESR}} \cdot \left(N^{\text{data,ESR}} - N_{\text{non-top}}^{\text{pred,ESR}} \right) \quad (7.12)$$

$$\alpha_{\text{top}}^{0j} = N_{\text{top}}^{\text{MC,SR,0j}} / N_{\text{top}}^{\text{MC,ESR,0j}} \quad (7.13)$$

where ϵ_0 is the jet veto efficiency. Figure 4.12 shows that, after pre-selection, top dominates the $e\mu/\mu e$ channels, but $Z/\gamma^* \to \ell\ell$ dominates the $ee/\mu\mu$ channels. For this reason, the top normalisation is derived from a combined $e\mu + \mu e$ channel and extrapolated to each $e\mu/\mu e/ee + \mu\mu$ SR.

The ESR is dominated by $t\bar{t}$ events featuring two b-jets. Since each b-quark originates from the decay of a top quark, the kinematic distributions of the two jets are similar. Thus, it is possible to suppress systematic uncertainties by correcting the MC using the second jet veto efficiency ϵ_1 measured in a top control region (CR).

7.3 Top

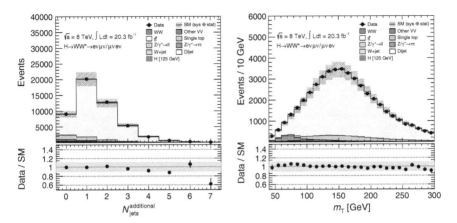

Fig. 7.7 The number of additional jets (*left*) and the m_T distribution (*right*) in the top control region used by the jet veto survival probability method. The fraction of events with zero additional jets is $\epsilon_{1,\text{top}}^{\text{data,CR}}$, as described in the text

The square of the data-driven correction to $\epsilon_{1,\text{top}}$ is used to correct the $\epsilon_{0,\text{top}}$ modelled by MC in the ESR:

$$\epsilon_{0,\text{top}}^{\text{pred,ESR}} = \epsilon_{0,\text{top}}^{\text{MC,ESR}} \left(\frac{\epsilon_1^{\text{data,CR}}}{\epsilon_{1,\text{top}}^{\text{MC,CR}}} \right)^2 . \qquad (7.14)$$

The top CR requires at least one b-tagged jet with $p_T > 25$ GeV. Then $\epsilon_1^{\text{data,CR}}$ is measured by counting the fraction of CR events with zero additional jets. These are defined as jets with $\Delta R > 1$ from the highest scoring b-tagged jet (see Fig. 7.7). It is found that (7.14) results in reduced uncertainties even when considering contributions from other top processes and jets other than the two b-jets.

The total uncertainties in the expected 0-jet top background is ~8%, dominated by theoretical uncertainties in the extrapolation α_{top}^{0j} and JES/JER uncertainties in the predicted jet veto efficiency $\epsilon_{0,\text{top}}^{\text{pred,ESR}}$.

7.3.2 1-jet Bin Estimation

In the 1-jet bin, the top background is suppressed by removing events with a b-tagged jet with $p_T > 20$ GeV. Since the b-tagging efficiency is associated with large uncertainties, the data-driven *jet b-tagging efficiency extrapolation method* is used to estimate this background.

An extended signal region (ESR) is defined by the pre-selection in Sect. 4.3.3. Since it is the $N_{\text{jets}} = 1$ selection and the b-tagged jet veto that introduce the largest

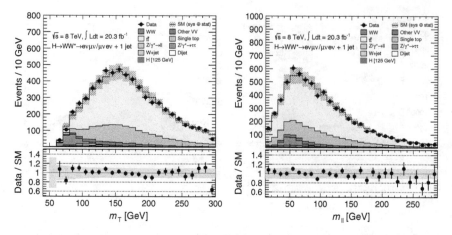

Fig. 7.8 The m_T (*left*) and $m_{\ell\ell}$ (*right*) distributions for events passing the pre-selection and containing 1 b-tagged jet. The jet b-tagging efficiency extrapolation method uses the top normalisation in this region and extrapolates to events containing 1 jet that is not b-tagged. Normalisation factors are applied

uncertainties to this background, the aim of the method is to estimate the number of events accepted by these cuts $N_{\text{top}}^{\text{pred,ESR,1j0}b}$, and then extrapolate to the 1-jet signal region (SR) with MC

$$N_{\text{top}}^{\text{pred,SR,1j0}b} = \alpha_{\text{top}}^{1j} \cdot N_{\text{top}}^{\text{pred,ESR,1j0}b} \tag{7.15}$$

$$\alpha_{\text{top}}^{1j} = N_{\text{top}}^{\text{MC,SR,1j0}b} / N_{\text{top}}^{\text{MC,ESR,1j0}b}. \tag{7.16}$$

The top normalisation is derived from a combined $e\mu + \mu e$ channel and extrapolated to each $e\mu/\mu e/ee + \mu\mu$ SR.

In order to estimate $N_{\text{top}}^{\text{pred,ESR,1j0}b}$, the number of 1-jet events in the ESR where the jet is b-tagged $N^{\text{data,ESR,1j1}b}$ is measured, which is highly pure in top events (see Fig. 7.8). Then the non-top contamination is subtracted, and the b-tagged jet selection is inverted using the expected b-tagging efficiency $\varepsilon_b^{\text{pred,ESR,1j}}$:

$$N_{\text{top}}^{\text{pred,SR,1j0}b} = \alpha_{\text{top}}^{1j} \cdot \frac{1 - \varepsilon_b^{\text{pred,ESR,1j}}}{\varepsilon_b^{\text{pred,ESR,1j}}} \cdot \left(N^{\text{data,ESR,1j1}b} \right) - N_{\text{non-top}}^{\text{pred,ESR,1j1}b}. \tag{7.17}$$

It should be emphasised that ε_b is a per-jet efficiency, rather than a per-event efficiency (cf. ϵ_0). Also, ε_b is the average b-tagging efficiency for a jet in a top event, and will include non-b-jets (i.e. it is a mixture of tagged b-jets and mis-tagged other jets).

The b-tagging efficiency of jets in top events ε_b is measured in a high-purity top control region (CR). These events pass the pre-selection (minus the $p_T^{\text{miss,corr}}$ cut) and feature two jets, at least one of which is b-tagged. ε_b is then measured using a tag-and-probe method (see Sect. 4.2.3), where the tag is a b-tagged jet and the probe is the other jet. Since ε_b is measured in 2-jet events but applied to 1-jet events, an MC-based correction is applied to the measured b-tagging efficiency

$$\varepsilon_b^{\text{pred,ESR,1j}} = \frac{\varepsilon_b^{\text{MC,ESR,1j}}}{\varepsilon_b^{\text{MC,CR,2j}}} \cdot \varepsilon_b^{\text{data,CR,2j}}. \quad (7.18)$$

The total uncertainty in the expected 1-jet top background is $\sim 7\,\%$, dominated by theoretical uncertainties in the MC-based correction to ε_b in (7.18).

7.3.3 $\geq$2-jet Bin Estimation

As in the 1-jet bin, a veto on b-tagged jets rejects the majority of the top background. However, even after this veto, top remains the largest background in the $\geq$2-jet bin. Fortunately, this enables a top control region (CR) to be defined which includes the b-tagged jet veto, greatly reducing the uncertainties due to b-tagging efficiencies.

The top CR is defined in a high-$m_{\ell\ell}$ region, similarly to the WW CRs in the 0-jet and 1-jet bins. It is defined by the same criteria as the $\geq$2-jet SR (minus the $m_{\tau\tau}$, $\Delta\phi\,(\ell,\ell)$ and VH cuts), but the $m_{\ell\ell}$ cut is changed to $m_{\ell\ell} > 80\,\text{GeV}$. The observed number of events is then extrapolated to the signal region (SR) using MC

$$N_{\text{top}}^{\text{pred,SR}} = \alpha_{\text{top}}^{\geq 2j} \left(N^{\text{data,CR}} - N_{\text{non-top}}^{\text{pred,CR}} \right) \quad (7.19)$$

$$\alpha_{\text{top}}^{\geq 2j} = N_{\text{top}}^{\text{MC,SR}} / N_{\text{top}}^{\text{MC,CR}}. \quad (7.20)$$

The uncertainty is dominated by theoretical uncertainties in the extrapolation $\alpha_{\text{top}}^{\geq 2j}$. Figure 7.9 shows the good description of experimental data in the CR, after application of the normalisation factors.

7.4 Z/γ^*

The Z/γ^* background is naturally split into $Z/\gamma^* \to \ell\ell$ and $Z/\gamma^* \to \tau\tau$ processes. The former contributes almost exclusively to the $ee/\mu\mu$ channels,[5] where it is the dominant background. The latter can contribute to any channel since the two $\tau \to \ell\nu_\ell\nu_\tau$ decays are independent, though is doubly suppressed by the small BR $(\tau \to \ell\nu_\ell\nu_\tau) = 17.6\,\%$ [5].

---

[5] In rare cases, $Z/\gamma^* \to \ell\ell$ events can enter the $e\mu/\mu e$ channels. For example $Z/\gamma^* \to \mu\mu\gamma$, where a muon radiates a photon which subsequently converts and is reconstructed as an electron.

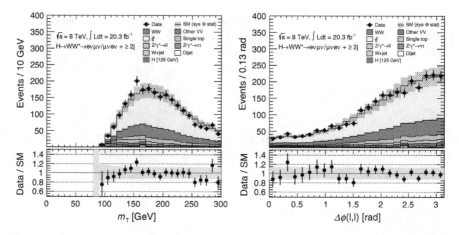

Fig. 7.9 The m_T (*left*) and $\Delta\phi\,(\ell,\ell)$ (*right*) distributions in the top control region of the $\geq$2-jet bin. Normalisation factors are applied

Although $Z/\gamma^* \to \ell\ell$ does not feature prompt neutrinos, degradation of the p_T^{inv} resolution due to high pile-up can cause some $Z/\gamma^* \to \ell\ell$ events to exhibit significant p_T^{inv}. Since $Z/\gamma^* \to \ell\ell$ is the dominant background in the $ee/\mu\mu$ channels, a high threshold of $E_{T,\text{rel}}^{\text{miss}} > 40\,\text{GeV}$ is used in the pre-selection, which is further tightened by tracker-based $p_{T,\text{rel}}^{\text{miss}}$ cuts. On the other hand, $Z/\gamma^* \to \tau\tau$ does feature prompt neutrinos and is suppressed by other cuts, such as the $m_{\tau\tau}$ veto.

The Z/γ^* backgrounds are estimated by data-driven techniques, in combination with MC modelling provided by ALPGEN+HERWIG. Overlap between the Z/γ^* MC and the $Z\gamma$ MC is removed through careful consideration of the MC event records.

7.4.1 Z/γ^* Boson Transverse Momentum

Selecting $ee/\mu\mu$ events with $\left|m_{\ell\ell} - m_Z\right| < 15\,\text{GeV}$ results in a very pure sample of $Z/\gamma^* \to \ell\ell$ events, enabling the MC to be validated. In doing so, the $p_{T,\ell\ell}$ distribution is found to be poorly modelled in the 0-jet bin for $p_{T,\ell\ell} > 30\,\text{GeV}$ (see Fig. 7.10), despite being well modelled inclusively. This is unsurprising: by vetoing events with jets whilst requiring a highly boosted Z/γ^* boson, we have selected a difficult phase space to model, sensitive to soft hadronic activity and jet shapes. It is important to model $p_{T,\ell\ell}$ accurately, as other observables such as $\Delta\phi\,(\ell,\ell)$ and $p_{T,\ell}^{\text{lead}}$ are correlated.

For this reason, a data-driven correction to the $p_{T,Z}$ distribution is employed. It is derived by comparing the observed and predicted $p_{T,\ell\ell}$ distributions in 0-jet $\mu\mu$ events with $\left|m_{\ell\ell} - m_Z\right| < 15\,\text{GeV}$. The $p_{T,Z}$ distribution in MC is then multiplied

7.4 Z/γ*

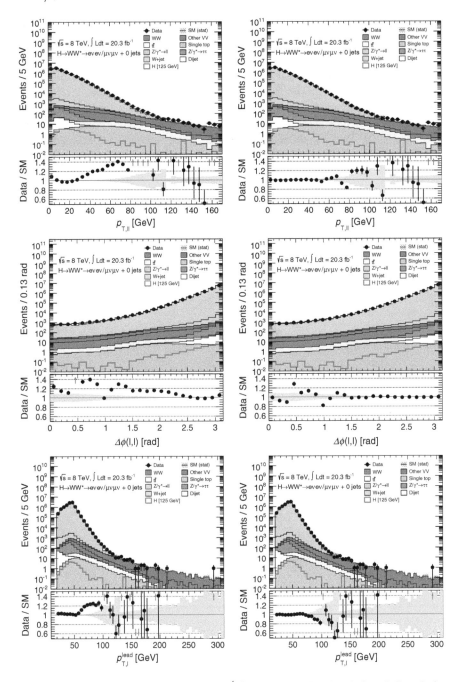

Fig. 7.10 Leptonic distributions in the 0-jet $Z/\gamma^* \to \ell\ell$ control region, before (*left*) and after (*right*) the $p_{T,Z}$ correction is applied

by this $p_{T,\ell\ell}$ correction. This is found to improve the modelling of detector-level observables such as $p_{T,\ell\ell}$, $\Delta\phi(\ell,\ell)$ and $p_{T,\ell}^{\text{lead}}$ (see Fig. 7.10). This correction is applied to the $Z/\gamma^* \to ee/\mu\mu/\tau\tau$ processes, though only in the 0-jet bin.

The $p_{T,Z}$ mismodelling might be different in the signal region to the Z control region where it is derived. The validity of this extrapolation of the correction was tested by deriving similar corrections to a SHERPA sample (instead of to experimental data). This indicated there is a correlation between the correction and the p_T^{inv} requirement. Thus, another data-driven correction is derived with an additional $p_T^{\text{miss,corr}} > 20\,\text{GeV}$ cut, which is used to estimate the uncertainty in the correction.

7.4.2 $Z/\gamma^* \to \tau\tau$ Estimation

The $Z/\gamma^* \to \tau\tau$ background is measured in a dedicated control region (CR), and then extrapolated to the signal region (SR) using MC

$$N_{Z\to\tau\tau}^{\text{pred,SR}} = \alpha_{Z\to\tau\tau} \cdot \left(N^{\text{data,CR}} - N_{\text{non-}Z\to\tau\tau}^{\text{pred,CR}}\right) \quad (7.21)$$

$$\alpha_{Z\to\tau\tau} = N_{Z\to\tau\tau}^{\text{MC,SR}} / N_{Z\to\tau\tau}^{\text{MC,CR}}. \quad (7.22)$$

CRs are defined in each jet bin of the $e\mu + \mu e$ channel, according to the event selection criteria in Table 7.2. Each $e\mu + \mu e$ CR is used to extrapolate to the corresponding jet-binned $e\mu/\mu e/ee + \mu\mu$ SRs. Figure 7.11 exhibits the excellent MC modelling of the shapes of this background, following application of the normalisation factors.

Table 7.2 Event selection criteria of the $Z/\gamma^* \to \tau\tau$ control regions

$e\mu/\mu e$		
$p_{T,\ell}^{\text{lead}} > 22$ and $p_{T,\ell}^{\text{sublead}} > 10$		
$m_{\ell\ell} > 12$		
$p_T^{\text{miss,corr}} > 20$		
0-jet bin	1-jet bin	$\geq$2-jet bin
–	$N_{b\text{-jets}} = 0$	$N_{b\text{-jets}} = 0$
–	$\max(m_{T,\ell}) > 50$	–
–	$m_{\tau\tau} > m_Z - 25$	–
–	–	Fail CJV or OLV
$m_{\ell\ell} < 80$	$m_{\ell\ell} < 80$	$m_{\ell\ell} < 70$
$\Delta\phi(\ell,\ell) > 2.8$	–	$\Delta\phi(\ell,\ell) > 2.8$

Cuts on energy, momentum and mass are given in GeV, and angular cuts are given in radians. The relevant observables are described in Chap. 4

7.4 Z/γ^*

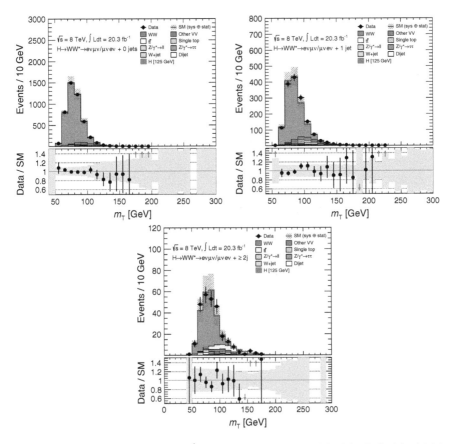

Fig. 7.11 The m_T distribution in the $Z/\gamma^* \to \tau\tau$ control regions of the 0-jet (*left*), 1-jet (*right*) and $\geq$2-jet (*bottom*) bins. Normalisation factors are applied

7.4.3 $Z/\gamma^* \to \ell\ell$ Estimation

$Z/\gamma^* \to \ell\ell$ is the dominant background to the $ee/\mu\mu$ channels, and this section will describe how it is estimated in the corresponding 0-jet and 1-jet bins (the $\geq$2-jet bin is not used in the $ee/\mu\mu$ channels).

$Z/\gamma^* \to \ell\ell$ is largely rejected by the Z boson mass veto, $\left|m_{\ell\ell} - m_Z\right| > 15\,\text{GeV}$, and the requirement of large missing transverse momentum, $E_{T,\text{rel}}^{\text{miss}} > 40\,(40)\,\text{GeV}$ and $p_{T,\text{rel}}^{\text{miss}} > 40\,(35)\,\text{GeV}$ in the 0-jet (1-jet) bin. Although the p_T^{inv} resolution is broadened by pile-up, for events to possess such high $E_{T,\text{rel}}^{\text{miss}}$ and $p_{T,\text{rel}}^{\text{miss}}$ suggests there is some mismeasured hadronic activity recoiling against the dilepton (+ jet) system. This situation is also enhanced by the $p_{T,\ell\ell} > 30\,\text{GeV}$ cut in the 0-jet bin.

This hadronic recoil must be soft and broad in order to avoid passing the jet reconstruction, and might not be modelled accurately. It is therefore preferable to

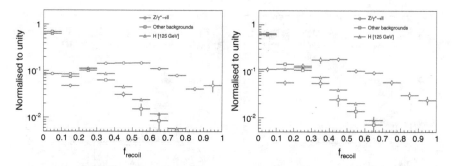

Fig. 7.12 f_{recoil} shape in the 0-jet (*left*) and 1-jet (*right*) signal regions of the $ee/\mu\mu$ channels, excluding the f_{recoil} cut itself. The $W+$ jet and dijet backgrounds are data-driven and all other processes are described by MC. *Error bars* show statistical uncertainties

use a data-driven estimation of this background. Fortunately, the f_{recoil} observable (described below) can discriminate $Z/\gamma^* \to \ell\ell$ from other processes in this phase space, and can be used to simultaneously suppress and estimate the $Z/\gamma^* \to \ell\ell$ background.

First, soft jets with $p_T > 10\,\text{GeV}$ and $|\eta| < 4.5$ are found, following the jet selection detailed in Sect. 4.2.5 minus the JVF cut. In the 0-jet bin, f_{recoil} is defined by

$$f_{\text{recoil}} = \left| \sum_{j \text{ in } \wedge} \text{JVF}_j \cdot \boldsymbol{p}_{T,j} \right| \Big/ p_{T,\ell\ell} \tag{7.23}$$

where $\wedge$ is the detector quadrant centred on $-\boldsymbol{p}_{T,\ell\ell}$. Approximately speaking, this is the fraction of $p_{T,\ell\ell}$ that can be balanced by soft hadronic activity in the opposing quadrant. In the 1-jet bin, it is the dilepton + jet system that must be balanced, and thus the quadrant $\wedge$ is centred upon $-\boldsymbol{p}_{T,\ell\ell j}$ and the denominator becomes $p_{T,\ell\ell j}$.

The tight p_T^{inv} criteria sculpt f_{recoil} in $Z/\gamma^* \to \ell\ell$ towards higher values, whereas the other background processes (and signal) feature high-p_T neutrinos and peak at lower values (see Fig. 7.12). A *template method* exploits the markedly different f_{recoil} shape of the $Z/\gamma^* \to \ell\ell$ background to estimate its contribution to the $ee+\mu\mu$ signal region.

In the template method, the signal region (SR) is defined by the full event selection of the $ee/\mu\mu$ channels, excluding the f_{recoil} cut (whose efficiency shall be predicted by the method). The f_{recoil} distribution $\mathcal{T}$ in the SR of the same-flavour (SF) channels, i.e. the $ee/\mu\mu$ channels, has two components with distinct shapes: $Z/\gamma^* \to \ell\ell$, and other processes (including signal). If the shape of both components are predicted, then their relative contributions can be fit using the observed f_{recoil} distribution in the SR

$$\mathcal{T}^{\text{data,SR,SF}} = K_{\text{non-}Z\to\ell\ell}^{\text{fit}} \cdot \mathcal{T}_{\text{non-}Z\to\ell\ell}^{\text{pred,SR,SF}} + K_{Z\to\ell\ell}^{\text{fit}} \cdot \mathcal{T}_{Z\to\ell\ell}^{\text{pred,SR,SF}} \tag{7.24}$$

where K_i^{fit} are prefactors determined by the fitting procedure, and the input distributions $\mathcal{T}_i^{\text{pred,SR,SF}}$ are determined by data-driven methods described below.

$\mathcal{T}_{\text{non-Z}\to\ell\ell}^{\text{pred,SR,SF}}$ is determined from the observed distribution in different-flavour (DF) events, i.e. the $e\mu/\mu e$ channels, $\mathcal{T}^{\text{data,SR,DF}}$. Note that this DF distribution is measured in the SR defined by the SF event selection criteria. In doing this, the normalisation factor $K_{\text{non-Z}\to\ell\ell}^{\text{fit}}$ becomes an extrapolation parameter $\alpha_{\text{non-Z}\to\ell\ell}^{\text{fit}}$ from DF to SF. This method is valid because the $Z/\gamma^* \to \ell\ell$ background to the DF channels is negligible, and the composition of the other processes is consistent between DF and SF events.

$\mathcal{T}_{Z\to\ell\ell}^{\text{pred,SR,SF}}$ is determined from the SF distribution measured in a $Z/\gamma^* \to \ell\ell$ control region (CR) $\mathcal{T}^{\text{data,CR,SF}}$. The CR is defined by the same selection criteria as the SR, except that the $m_{\ell\ell} < 55\,\text{GeV}$ criteria becomes $\left|m_{\ell\ell} - m_Z\right| < 15\,\text{GeV}$. As the p_T^{inv} and $p_{T,\ell\ell}$ criteria remain in the CR definition, f_{recoil} is sculpted similarly to the SR. However, this also means that the contribution of other processes in the CR is non-negligible and must be subtracted. This contamination is estimated from the distribution measured in DF events passing the same CR selection $\mathcal{T}^{\text{data,CR,DF}}$, with the extrapolation from DF to SF predicted by dedicated data-driven methods described elsewhere or by MC.

Thus, (7.24) becomes

$$\mathcal{T}^{\text{data,SR,SF}} = \alpha_{\text{non-Z}\to\ell\ell}^{\text{fit}} \cdot \mathcal{T}^{\text{data,SR,DF}} + \alpha_{Z\to\ell\ell}^{\text{fit}} \cdot \left(\mathcal{T}^{\text{data,CR,SF}} - \alpha_{\text{non-Z}\to\ell\ell}^{\text{pred}} \cdot \mathcal{T}^{\text{data,CR,DF}}\right) \tag{7.25}$$

where $\alpha_{\text{non-Z}\to\ell\ell}^{\text{fit}}$ and $\alpha_{\text{non-Z}\to\ell\ell}^{\text{pred}}$ are extrapolations from DF to SF, and $\alpha_{Z\to\ell\ell}^{\text{fit}}$ is an extrapolation in $m_{\ell\ell}$. Note that the α_i are prefactors and leave template shapes unchanged. The sensitivity of this method to the signal strength is tested and found to be negligible.

The final event selection criterion in the $ee/\mu\mu$ channels is $f_{\text{recoil}} < 0.1$. The efficiency of this cut for both $Z/\gamma^* \to \ell\ell$ and other processes is inherently predicted by the template method. To simplify matters, the templates are reduced to two bins: $f_{\text{recoil}} < 0.1$ and $f_{\text{recoil}} > 0.1$. With two bins and two free parameters α_i^{fit}, the system can be solved exactly: instead of fitting the f_{recoil} shape, the method estimates the efficiency of the f_{recoil} cut. The dominant uncertainty in the 0-jet $Z/\gamma^* \to \ell\ell$ estimation arises from correlations between f_{recoil} and $m_{\ell\ell}$, and is evaluated with MC to be 32%. The 1-jet bin is dominated by statistical uncertainties.

7.5 Summary of Normalisation Factors

It is helpful to express the data-driven estimations of several backgrounds in terms of a normalisation factor, which is the ratio of the data-driven predicted yield to the MC-only predicted yield. Their pre-fit and post-fit values are summarised in

Table 7.3 The data-driven normalisation factor used to scale the MC description of each background process

Process	Pre-fit (post-fit) normalisation factor		
	0-jet	1-jet	≥2-jet
WW	1.22 (1.22)	1.06 (1.12)	–
Non-WW diboson	0.91 (0.95)	0.95 (0.84)	–
Top	1.09 (1.09)	1.04 (1.03)	1.00 (1.00)
$Z/\gamma^* \to \tau\tau$	1.00 (0.99)	1.05 (1.05)	0.96 (0.97)

These can change in the fit due to the pull of nuisance parameters (see Sect. 8.2.2). The $W+$ jet and dijet processes are fully data-driven

Table 7.3, where the fit can change the values due to the pull of nuisance parameters (see Sect. 8.2.2).

These can also be used to improve the background estimations in regions other than the signal region; however, this does neglect theoretical uncertainties in the extrapolation from the corresponding control region. Plotted distributions in Chaps. 4–7 use pre-fit normalisation factors in this way.

References

1. ATLAS Collaboration, Background estimation in the $H \to WW^{(*)} \to \ell\nu\ell\nu$ analysis with 20.7 fb^{-1} of data collected with the ATLAS detector at $\sqrt{s} = 8$ TeV, ATL-COM-PHYS-2013-1630 (2013) (ATLAS internal)
2. ATLAS Collaboration, Measurements of Higgs boson production and couplings in diboson final states with the ATLAS detector at the LHC. Phys. Lett. B **726**, 88 (2013). arXiv:1307.1427 [hep-ex]
3. J. Alwall, Private communication (2012)
4. F. Krauss, F. Siegert, Private communication (2012)
5. Particle Data Group, Review of particle physics. Phys. Rev. D **86**, 010001 (2012), and 2013 partial update for the 2014 edn

Chapter 8
Experimental Results

Chapters 4–7 describe the details of the $H \to WW$ analysis: the selection of signal events and the rejection of backgrounds, and how each process is modelled. By comparing the expected and observed events passing the selection criteria, with a careful treatment of statistical and systematic uncertainties, it is possible to make statistically meaningful statements about the $H \to WW$ process, based upon the Run I dataset.

Section 8.1 briefly summarises the experimental and theoretical sources of systematic uncertainty. Then, Sect. 8.2 describes the statistical treatment used to extract the experimental results, which are finally presented in Sect. 8.3.

8.1 Systematic Uncertainties

In addition to statistical uncertainties in the observed number of events and the expected number of events (due to finite MC sample sizes), multiple sources of systematic uncertainty should be considered. Many of these were introduced throughout Chaps. 4–7, but a short summary is presented here.

8.1.1 Experimental Uncertainties

Experimental uncertainties arise due to a mismodelling of the detector performance, particularly in the reconstruction efficiency and energy calibration of physics objects. These introduce uncertainties either directly into the expected acceptance (as with signal) or via the extrapolation from a control region (as with many backgrounds).

Trigger efficiency
The lepton trigger efficiencies and their uncertainties are measured as a function of p_T and η using tag-and-probe of $Z \to \ell\ell$ events (see Sect. 4.3.2). This includes the efficiency of matching the trigger to the lepton object.

Lepton reconstruction efficiency

The lepton selection efficiencies and their uncertainties are measured as a function of p_T and η using tag-and-probe of $Z \to \ell\ell$ events (see Sects. 4.2.3 and 4.2.4). This is performed separately for each step in the lepton selection, i.e. reconstruction, identification, isolation and primary vertex association. The efficiency uncertainties are $<0.5\,\%$ for muons and $<3\,\%$ for electrons.

Lepton energy scale and resolution

Lepton energy scales and resolutions are calibrated in situ as a function of p_T and η using $J/\psi \to \ell\ell$ and $Z \to \ell\ell$ resonance events, and the associated uncertainties are also derived during this calibration. Scale uncertainties are $<0.5\,\%$ and their impact is assessed by varying lepton energies by $\pm 1\sigma$. Resolution uncertainties are $<1\,\%$ and their impact is assessed by varying the smearing of lepton energies by $\pm 1\sigma$.

Jet reconstruction efficiency

The jet selection criteria feature a cut upon the jet vertex fraction (JVF) in order to reduce pile-up jets (see Sect. 4.2.5). Unlike the trigger, lepton reconstruction and b-tagging efficiencies, the JVF efficiency is not subject to a data-driven scale factor and so MC mismodelling introduces a systematic uncertainty to the N_{jets} distribution. The mismodelling of the JVF efficiency is evaluated as $1.2\,\%$ in a $Z/\gamma^* \to \ell\ell$ control region, resulting in negligible jet bin migrations. The data-driven background estimations are jet-binned and mostly avoid this issue.

Jet energy scale and resolution

The jet energy scale (JES) is calibrated as a function of p_T and η, as described in Sect. 4.2.5. The in situ calibration method of balancing jets against well-measured reference objects introduces experimental and theoretical uncertainties to the JES. Also, uncertainties in the pile-up environment introduce uncertainties to the JES through the pile-up subtraction step of the calibration. Finally, the JES is affected by uncertainties in the jet flavour composition and the detector response to each flavour. The uncertainty components are treated separately (the total uncertainty is $<7\,\%$) and their impact is assessed by varying jet energies by $\pm 1\sigma$.

The jet energy resolution (JER) and its uncertainty are measured in situ as a function of p_T and η using dijet events [1]. The bisector method projects the dijet p_T imbalance in the direction bisecting the two jets and in the orthogonal direction. At hadron-level the two components are expected to have the same variance, but at detector-level the variance of the latter is increased by the JER. Thus, measurements of the two components allow the JER to be evaluated. The uncertainty is $<10\,\%$ and its impact is assessed by increasing the smearing of jet energies by $+1\sigma$.

Jet b-tagging efficiency

The b-tagging efficiency for b-jets is calibrated as a function of jet p_T, using dileptonic $t\bar{t}$ decays in a combinational likelihood method [2]. At $p_T \approx 20$ GeV the measurement is limited by JES uncertainties, but at $p_T > 40$ GeV it is limited

8.1 Systematic Uncertainties

by modelling uncertainties in the jet flavour composition. Uncertainties in the b-tagging efficiency of c-jets and light jets are also considered, though have little effect.

p_T^{inv} modelling

As described in Sect. 4.2.7, E_T^{miss}, p_T^{miss} and $p_T^{miss,corr}$ are defined using the calibrated electron, muon and jet objects, and are therefore correlated to the electron, muon and jet energy scales whose uncertainties are outlined above. In the E_T^{miss} calculation, uncertainties in the scale of the unassociated soft calorimeter deposits are considered. In the p_T^{miss} and $p_T^{miss,corr}$ calculations, uncertainties in the p_T imbalance of tracks unassociated with a physics object are considered (e.g. due to the presence of neutral particles).

Lepton fake factors

The lepton fake factors f_ℓ are important to the $W+$ jet and dijet background estimations, and are detailed in Sect. 7.1. f_ℓ^{W+jet} is measured in $Z+$ jet events, and is dominated by statistical uncertainties, uncertainties in the electroweak background subtraction and theoretical uncertainties in MC-based corrections. However, a component of the uncertainties cancels in the same-sign control region method (see Sect. 7.2.1). f_ℓ^{dijet} is measured in dijet events and is dominated by theoretical uncertainties in MC-based corrections.

Photon conversions

Modelling photon conversions is important to the $W\gamma$ background estimation, in particular the efficiency of two electron identification criteria: the hit in the first pixel layer and the conversion vertex reconstruction. Mismodelling of these criteria is tested in $Z\gamma$ events and applied as a systematic uncertainty (see Sect. 7.2.2).

Pile-up

Uncertainties in the pile-up environment affect the jet calibration, as outlined above. Pile-up can also introduce additional hard jets that lead to migrations between jet bins. Uncertainties in pile-up lead to migration uncertainties of 0.5 % and 1.0 % in the 0-jet and 1-jet bins respectively, and are neglected as they are much smaller than the ggF jet binning uncertainties.

The effect of out-of-time pile-up upon detector electronics might also be mismodelled. This is found to be negligible.

Luminosity

The luminosity measurement is calibrated during van der Meer scans (see Sect. 3.2.1), with a relative uncertainty of 2.8 % for the $\sqrt{s} = 8$ TeV dataset.

8.1.2 Theoretical Uncertainties

Many theoretical uncertainties are also considered (and indeed are included within the experimental uncertainties above). Theoretical uncertainties in the expected signal are described in Chap. 5. Theoretical uncertainties in the expected background appear when some aspect of a data-driven technique relies upon MC modelling. Some of the uncertainties considered are presented in Chaps. 6 and 7.

The sources of uncertainty usually considered are: higher order corrections in the perturbative series, incoming parton distribution functions, and aspects of the MC modelling (e.g. hadronisation and underlying event models).

8.2 Statistical Model

A statistical model is built within the HISTFACTORY software package [3], incorporating the various data-driven techniques described in Chaps. 6 and 7. The experimental data and MC-modelled expectations found within each signal and control region allow the model to be constrained, and for various hypotheses to be tested.

8.2.1 Discriminant Observables

The fitting procedure discriminates between the signal and background processes through the distribution of certain sensitive observables. However, the degree to which this information can be exploited is limited by statistical uncertainties. The sensitivity of the fitting procedure can be optimised through the choice of discriminating observables, the number of bins in each observable, and the location of the bin boundaries.

The analysis features eight signal regions, defined by the criteria in Sect. 4.3: $\{e\mu, \mu e, ee + \mu\mu\} \otimes \{\text{0-jet,1-jet}\} \oplus \{e\mu, \mu e\} \otimes \{\geq \text{2-jet}\}$. A three-dimensional fit of m_T, $p_{T,\ell}^{\text{sublead}}$ and $m_{\ell\ell}$ is performed in the $\{e\mu, \mu e\} \otimes \{\text{0-jet,1-jet}\}$ signal regions, whereas a one-dimensional m_T fit is used in the others.

Additionally, in the $\{e\mu, \mu e, ee + \mu\mu\} \otimes \{\text{0-jet,1-jet}\}$ signal regions, the m_T bin boundaries are chosen such that there are an equal number of expected signal events in each bin. This is applied independently within each $p_{T,\ell}^{\text{sublead}}$-$m_{\ell\ell}$ bin in the $e\mu/\mu e$ channels. In the $\geq$2-jet bin, fixed m_T bin boundaries are used.

The binning of each discriminant observable is shown in Table 8.1, and the inclusive m_T distribution of each jet bin is shown in Fig. 8.1.

8.2.2 Likelihood Function

The statistical model describing the experiment depends upon a set of parameters $\alpha = (\mu, \theta)$, where $\mu = \sigma/\sigma_{\text{SM}}$ is the signal strength (a parameter of interest) and θ is the set of nuisance parameters (e.g. trigger efficiency, WW extrapolation parameter α_{WW}). The Higgs boson mass m_H could also be treated as a parameter of interest, but hypothesis testing is usually performed as a raster scan of m_H in practice.

8.2 Statistical Model

Table 8.1 The binning of each discriminant observable for each signal region

N_{jets}	Channel	Bin boundaries (GeV)			N_{bins}
		$p_{\text{T},\ell}^{\text{sublead}}$	$m_{\ell\ell}$	m_{T}	
0-jet	$e\mu, \mu e$	[10, 15, 20, ∞]	[10, 30, 55]	10 bins	60
0-jet	$ee + \mu\mu$	[10, ∞]	[12, 55]	10 bins	10
1-jet	$e\mu, \mu e$	[10, 15, 20, ∞]	[10, 30, 55]	6 bins	36
1-jet	$ee + \mu\mu$	[10,∞]	[12, 55]	6 bins	6
≥2-jet	$e\mu, \mu e$	[10, ∞]	[10, 55]	[0, 50, 80, 130, ∞]	4

The m_{T} bin boundaries in the 0-jet and 1-jet bins are chosen to produce a flat signal

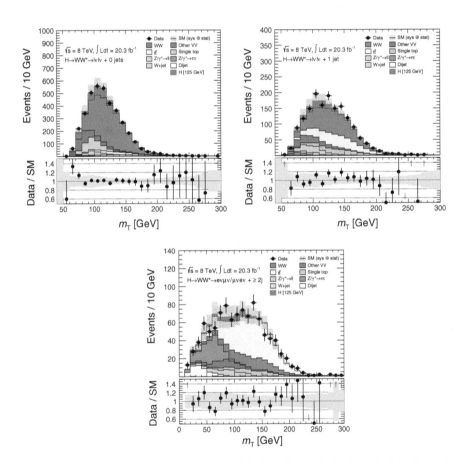

Fig. 8.1 The m_{T} distribution in the signal region of the 0-jet (*left*), 1-jet (*right*) and ≥2-jet (*bottom*) bin. Normalisation factors are applied

The likelihood function expresses how likely a set of parameter values are, given that a particular dataset is observed. It is defined as the probability of producing the observed dataset, when the parameter values are input into the statistical model,

$$\mathcal{L}(\mu, \boldsymbol{\theta}) = \mathcal{L}(\mu, \boldsymbol{\theta} \mid \mathcal{D}^{\text{SR}}, \mathcal{D}^{\text{CR}}) \qquad (8.1)$$
$$= f(\mathcal{D}^{\text{SR}}, \mathcal{D}^{\text{CR}}; \mu, \boldsymbol{\theta}) \qquad (8.2)$$

where $\mathcal{D}^{\text{SR}}$ and $\mathcal{D}^{\text{CR}}$ are the observed datasets in the signal regions (SRs) and control regions (CRs), respectively. This can be decomposed into a product of probabilities

$$\mathcal{L}(\mu, \boldsymbol{\theta}) = \prod_{c \in \text{channels}} f(\mathcal{D}_c^{\text{SR}}; \mu, \boldsymbol{\theta}) \prod_{c \in \text{channels}} f(\mathcal{D}_c^{\text{CR}}; \mu, \boldsymbol{\theta}) \prod_{i \in \text{n.p.}} f_i(\tilde{\theta}_i; \theta_i, \Delta\tilde{\theta}_i) \qquad (8.3)$$

where the SR channels are the eight SRs described in Sect. 8.2.1, and the CR channels include the data used in data-driven background estimations. Each $f_i(\tilde{\theta}_i; \theta_i, \Delta\tilde{\theta}_i)$ acts to constrain the corresponding nuisance parameter, and is often a Gaussian distribution. The nominal value $\tilde{\theta}$ and uncertainty $\Delta\tilde{\theta}$ are determined by an auxiliary measurement, either experimental (e.g. trigger efficiency) or theoretical (e.g. α_{WW}).

The probability of producing an observed distribution $\mathcal{D}_c$ is described by a product of Poisson distributions

$$f(\mathcal{D}_c; \mu, \boldsymbol{\theta}) = \prod_{b \in \text{bins}} \text{Pois}\left(N_{cb}^{\text{obs}}; \mu N_{cb}^{\text{sig}}(\boldsymbol{\theta}) + N_{cb}^{\text{bkg}}(\boldsymbol{\theta})\right) \qquad (8.4)$$

where $\text{Pois}(x; \lambda) = \lambda^x e^{-\lambda}/x!$, N_{cb}^{obs} is the number of observed events, $N_{cb}^{\text{sig}}(\boldsymbol{\theta})$ is the number of expected signal events for a Standard Model Higgs boson (estimated by MC), and $N_{cb}^{\text{bkg}}(\boldsymbol{\theta})$ is the number of expected background events (estimated by data-driven techniques embedded in the statistical model). The product is over the bins of the distribution $\mathcal{D}_c$, which is one- or three-dimensional in the case of the SRs, and zero-dimensional in many of the CRs (i.e. a single bin).

The expected numbers of events depend upon the nuisance parameters according to

$$N_{cb}^{\text{sig}}(\boldsymbol{\theta}) = N_{cb}^{\text{sig}}(\tilde{\boldsymbol{\theta}}) \prod_{i \in \text{n.p.}} \nu_{cbi}(\theta_i - \tilde{\theta}_i) \qquad (8.5)$$

$$N_{cb}^{\text{bkg}}(\boldsymbol{\theta}) = \sum_{p \in \text{processes}} N_{cbp}^{\text{bkg}}(\tilde{\boldsymbol{\theta}}) \prod_{i \in \text{n.p.}} \nu_{cbpi}(\theta_i - \tilde{\theta}_i) \qquad (8.6)$$

where $N(\tilde{\boldsymbol{\theta}})$ is the expected events with the nominal nuisance parameters as determined by auxiliary measurements, and $\nu_i(\theta_i - \tilde{\theta}_i)$ factorises the dependence upon the nuisance parameter θ_i. The $\nu_i(\theta_i - \tilde{\theta}_i)$ are often determined by evaluating $\nu_i(+\Delta\tilde{\theta}_i)$ and $\nu_i(-\Delta\tilde{\theta}_i)$, and then interpolating and extrapolating. Some nuisance parameters affect all processes (e.g. trigger efficiency), whilst others only affect a particular process (e.g. α_{WW}). Correlations can exist between bins and channels, e.g. the $\nu_i(\theta_i - \tilde{\theta}_i)$ for a normalisation parameter would be 100 % correlated between bins. Statistical uncertainties in the MC samples are implemented as uncorrelated uncertainties in the total expected events in each bin, where the corresponding constraint $f_i(\tilde{\theta}_i; \theta_i, \Delta\tilde{\theta}_i)$ is a Poisson distribution.

8.2.3 Hypothesis Testing

The statistical techniques of hypothesis testing employed by Higgs boson searches at the LHC are fully described in references [4, 5]. Those used in the $H \rightarrow WW$ search to exclude, discover and measure the Higgs boson are briefly summarised below.

Hypothesis testing is based around the profile likelihood ratio, defined as

$$\lambda(\mu) = \frac{\mathcal{L}(\mu, \hat{\hat{\theta}}(\mu))}{\mathcal{L}(\hat{\mu}, \hat{\theta})} \quad (8.7)$$

where $\hat{\mu}$ and $\hat{\theta}$ are the parameter values that maximise the likelihood, and $\hat{\hat{\theta}}(\mu)$ are the nuisance parameter values that maximise the likelihood when μ is fixed. Maximal agreement with experimental data is expressed by the maximum $\lambda(\hat{\mu}) = 1$, and statistical and systematic uncertainties determine the shape of $\lambda(\mu)$. The maximised $\hat{\theta}$ and $\hat{\hat{\theta}}(\mu)$ can be pulled, and constrained differently, with respect to the auxiliary measurements $\tilde{\theta}$. An alternative version of the profile likelihood ratio is defined as

$$\tilde{\lambda}(\mu) = \begin{cases} \frac{\mathcal{L}(\mu, \hat{\hat{\theta}}(\mu))}{\mathcal{L}(\hat{\mu}, \hat{\theta})} & \text{if } \hat{\mu} \geq 0 \\ \frac{\mathcal{L}(\mu, \hat{\hat{\theta}}(\mu))}{\mathcal{L}(0, \hat{\hat{\theta}}(0))} & \text{if } \hat{\mu} < 0 \end{cases} \quad (8.8)$$

which limits the model to $\hat{\mu} \geq 0$.

Frequentist techniques are used to test three different scenarios. First, we attempt to exclude a Standard Model (i.e. $\mu = 1$) Higgs boson at the 95 % confidence level (CL), by testing $\mu < 1$. Second, we attempt to discover a Higgs boson (i.e. $\mu \neq 0$) with $\geq 5\sigma$ significance, by testing $\mu > 0$. Third, we make a measurement of the signal strength μ (with a two-sided confidence interval). Each scenario is tested as a raster scan of m_H.

Under each scenario (detailed below), a test statistic is constructed to quantify the disagreement between the observed data and a null hypothesis. A p-value is then computed, which is the probability of observing a test statistic at least as extreme as that observed, under the assumption of the null hypothesis. The p-value is computed by integrating a sampling distribution of the test statistic, which is based upon the likelihood function. This is done using an asymptotic formula, which is valid in the large-sample limit, though an ensemble of pseudoexperiments could be used instead [4]. The validity of the asymptotic formula for this analysis has previously been checked against pseudoexperiments.

Finally, the median confidence levels expected under the assumption of an alternative hypothesis are calculated. This is done using a single pseudoexperiment, known as the Asimov dataset, corresponding to the *exact* expectation (i.e. pseudodata with fractional numbers of events) [4].

Exclusion

To determine an upper limit on the signal strength, a one-sided test statistic is constructed under the assumption of a signal-plus-background null hypothesis:

$$\tilde{q}_\mu = \begin{cases} -2\ln \tilde{\lambda}(\mu) & \text{if } \hat{\mu} \leq \mu \\ 0 & \text{if } \hat{\mu} > \mu \end{cases} \quad (8.9)$$

where use of $\tilde{\lambda}(\mu)$ ensures that only $\mu \geq 0$ is considered physical. The compatibility of the data with the null hypothesis is quantified by the p-value

$$p_\mu = \int_{\tilde{q}_{\mu,\text{obs}}}^{\infty} f(\tilde{q}_\mu \mid \mu, \hat{\hat{\theta}}(\mu)) \, d\tilde{q}_\mu \quad (8.10)$$

where $f(\tilde{q}_\mu \mid \mu, \hat{\hat{\theta}}(\mu))$ is the sampling distribution of the test statistic under the assumption of an underlying signal strength μ, and $\tilde{q}_{\mu,\text{obs}}$ is the observed test statistic under the same assumption. Thus, a 95 % CL upper limit upon μ is set at the value which satisfies $p_\mu = 0.05$.

An unfortunate feature of this method is that a downward fluctuation in the observed data can exclude models where little sensitivity is expected. For this reason the modified frequentist CL_s technique is used [6, 7], which defines the p-value

$$p'_\mu = \frac{p_\mu}{1 - p_b} \quad (8.11)$$

where incompatibilities with the null hypothesis are down-weighted if the data is also incompatible with the background-only hypothesis, through

$$p_b = \int_{-\infty}^{\tilde{q}_{\mu,\text{obs}}} f(\tilde{q}_\mu \mid 0, \hat{\hat{\theta}}(0)) \, d\tilde{q}_\mu . \quad (8.12)$$

Thus, a 95 % CL upper limit upon μ is set at the value which satisfies $p'_\mu = 0.05$. The Standard Model Higgs boson is considered excluded if $\mu < 1$ at the 95 % CL. The exclusion sensitivity is evaluated by calculating the expected μ upper limit under the assumption of a background-only alternative hypothesis.

8.2 Statistical Model

Discovery

For discovery, a one-sided test statistic is constructed under the assumption of a background-only null hypothesis:

$$q_0 = \begin{cases} -2\ln\lambda(0) & \text{if } \hat{\mu} \geq 0 \\ 0 & \text{if } \hat{\mu} < 0. \end{cases} \quad (8.13)$$

The compatibility of the data with the null hypothesis is quantified by the p-value

$$p_0 = \int_{q_{0,\text{obs}}}^{\infty} f(q_0 \mid 0, \hat{\hat{\theta}}(0))\, dq_0 . \quad (8.14)$$

The Higgs boson is considered discovered if the background-only hypothesis is excluded with $p_0 < 2.87 \times 10^{-7}$, corresponding to a significance of at least five standard deviations (i.e. $\geq 5\sigma$) in a Gaussian distribution. The discovery sensitivity is evaluated by calculating the expected p_0 under the assumption of a signal-plus-background alternative hypothesis, with $\mu = 1$. As seen in Sect. 8.3, an m_H value must also be chosen for this hypothesis.

Measurement

For the μ measurement, a two-sided test statistic is constructed under the assumption of a signal-plus-background null hypothesis:

$$t_\mu = -2\ln\lambda(\mu) . \quad (8.15)$$

The compatibility of the data with the null hypothesis is quantified by the p-value

$$p_\mu = \int_{t_{\mu,\text{obs}}}^{\infty} f(t_\mu \mid \mu, \hat{\hat{\theta}}(\mu))\, dt_\mu . \quad (8.16)$$

The nominal μ value will be $\hat{\mu}$, and the 68 % CL confidence interval is set by the values which satisfy $p_\mu = 0.16$. The measurement sensitivity is evaluated by calculating the expected $\Delta\mu$ under the assumption of a signal-plus-background hypothesis, with $\mu = 1$. Again, an m_H value must be chosen for this hypothesis.

8.3 Results

The entire LHC Run I dataset was analysed, corresponding to an integrated luminosity of 4.5 fb^{-1} at $\sqrt{s} = 7$ TeV and 20.3 fb^{-1} at $\sqrt{s} = 8$ TeV. The major differences of the 7 TeV analysis, with respect to Chap. 4, are the absence of the $\geq$2-jet bin and the dilepton triggers, and the use of a dijet fake factor in the $W +$ jet background estimation. Additionally, the number of m_T bins used in the fit is reduced by a factor of two. Some minor differences also exist in the object and event selection criteria due to the less harsh pile-up environment at 7 TeV (see Sect. 3.2.2).

8.3.1 Exclusion, Discovery and Measurement of $gg \to H \to WW$

The event selection criteria of the $gg \to H \to WW$ analysis are chosen such that the dominant production mode of the selected Higgs boson events is ggF. Nevertheless, other production modes contribute to the signal acceptance. This effect is small in the 0-jet and 1-jet bins, but VBF (VH) contributes 15 % (10 %) of signal events to the $\geq$2-jet bin. Such events are treated as signal in the fit, and their expected yield is scaled by the same μ as ggF. This has a small effect upon the results, and splitting μ according to production mode shall be revisited in Sect. 8.3.2.

The observed and expected (under a background-only hypothesis) 95 % CL upper limit on μ is shown as a function of the mass under test m_H in Fig. 8.2a. The mass range where this limit is below unity is excluded at the 95 % CL, when considering an SM Higgs boson ($\mu = 1$). In the absence of a Higgs boson, the expected excluded region is 116 to 200 GeV.[1] However, the observed excluded region is 132 to 200 GeV. The fact that the observed exclusion is weaker than expected indicates an excess of events consistent with a Higgs boson. Since the mass resolution of the $H \to WW$ analysis is poor, the impact upon the exclusion is broad in m_H.

To quantify the significance of the excess of events, Fig. 8.2b shows the observed p_0 as a function of m_H. The maximum observed significance is 4.8σ, which occurs when testing $m_H = 130$ GeV, though the poor mass resolution leads to a relatively broad p_0 curve. This is very strong evidence for the existence of the Higgs boson, though does not pass the 5σ criterion for "discovery".

The best-fit signal strength $\hat{\mu}$ is shown as a function of the m_H under test in Fig. 8.2c. A range of Higgs boson masses are consistent with the assumption that the Higgs boson behaves as predicted by the SM (i.e. $\mu = 1$). The data are also consistent with a lower mass Higgs boson with $\mu > 1$ or a higher mass Higgs boson with $\mu < 1$.

8.3.2 Combination with VBF Analysis

The ggF analysis described in this thesis is combined with a multivariate VBF analysis, which selects events featuring two high-p_T jets separated by a large rapidity gap devoid of hadronic activity. This analysis features a higher signal-to-background ratio than the ggF analysis, but has a smaller expected number of events. It was briefly

---

[1] This analysis is optimised to search for a low-mass Higgs boson and therefore the region $m_H > 200$ GeV is not considered. A dedicated high-mass search for $H \to WW$ is described in reference [8].

8.3 Results

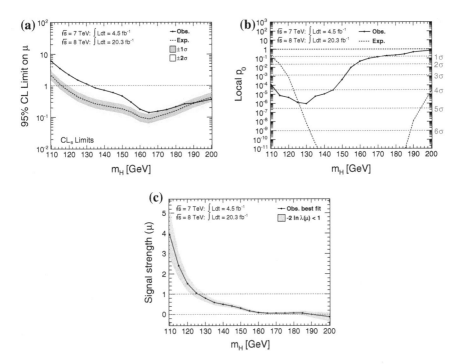

Fig. 8.2 Results of the ggF analysis. **a** The observed (*solid*) 95% CL upper limit on the signal strength μ as a function of the mass under test m_H, and the expectation (*dashed*) under the background-only hypothesis. **b** The observed (*solid*) p_0 as a function of m_H and the expectation (*dashed*) under the signal-plus-background hypothesis with $\mu = 1$ and a hypothesised mass equal to that under test. **c** The best-fit signal strength $\hat{\mu}$ (*solid*) as a function of m_H, with the 68% CL interval shown (*blue band*)

introduced in Sect. 4.3.7, because the ≥2-jet bin of the ggF analysis vetoes events selected by the VBF analysis in order to maintain orthogonality. The VBF analysis will be fully described in the upcoming paper [9].

The 95% CL upper limit on μ is shown as a function of m_H in Fig. 8.3a. The expected (in the absence of a Higgs boson) excluded region is 114 to 200 GeV, exhibiting only a small improvement upon the limit from the ggF analysis. The observed excluded region is 132 to 200 GeV.

The observed p_0 is shown in Fig. 8.3b as a function of the mass under test m_H. The maximum observed significance is 6.1σ, occurring when testing $m_H = 130$ GeV. This constitutes a discovery of the Higgs boson, made by searching for the ggF and VBF production modes and the WW decay channel.

The best-fit signal strength $\hat{\mu}$ is shown as a function of the m_H under test in Fig. 8.3c. By including m_H as a parameter of interest in the fit, it is possible to test which (μ, m_H) pair are most favoured by the data. This is presented in Fig. 8.3d, together with likelihood contours. The best fit values are $(\hat{\mu}, \hat{m}_H) = (0.90, 128\,\text{GeV})$.

These results exhibit more than 5σ significance for $H \rightarrow WW \rightarrow \ell\nu\ell\nu$, and consequently for the existence of the Higgs boson itself. LHC searches for $H \rightarrow \gamma\gamma$ and $H \rightarrow ZZ$ have found consistent evidence, which is summarised in Sect. 9.1.

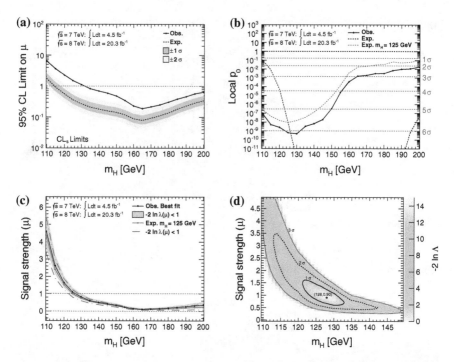

Fig. 8.3 Results of the combined ggF+VBF analysis. **a** The observed (*solid*) 95 % CL upper limit on the signal strength μ as a function of the mass under test m_H, and the expectation (*dashed*) under the background-only hypothesis. **b** The observed (*solid*) p_0 as a function of m_H and the expectation (*black dashed*) under the signal-plus-background hypothesis with $\mu = 1$ and a hypothesised mass equal to that under test. The expectation (*blue dashed*) for a hypothesised mass $m_H = 125$ GeV is also shown. **c** The best-fit signal strength $\hat{\mu}$ (solid) as a function of m_H, with the 68 % CL interval shown (*blue band*). The expectation (*red*) for a hypothesised mass $m_H = 125$ GeV is also shown. **d** The best-fit signal strength $\hat{\mu}$ and mass $\hat{m}_H$ (*marker*), with the likelihood contours also shown

These two channels yield much better mass sensitivity, and observe $m_H \approx 125$ GeV. For this reason, the expected p_0 and $\hat{\mu}$ under the assumption of a SM Higgs boson with $m_H = 125$ GeV are also included in Figs. 8.3b, c respectively. It can be seen that the observed p_0 is smaller than expected (i.e. the significance of the incompatibility with the null hypothesis is greater), though the shape remains consistent. The best-fit signal strength at $m_H = 125$ GeV is observed to be

$$\hat{\mu} = 1.11 \pm 0.22.$$

The excess of events is further characterised in Fig. 8.4. In comparison, the combined $H \to WW$ analysis performed by the CMS experiment measures $\hat{\mu} = 0.72 \pm 0.19$ at $m_H = 125.6$ GeV [10].

It is possible to assign the ggF and VBF production modes independent signal strength parameters, with μ_{ggF} constrained by the ggF analysis and μ_{VBF} constrained by the VBF analysis. This is interesting because they are dominated by different couplings, and the loop in the ggF process can be sensitive to undiscovered massive

8.3 Results

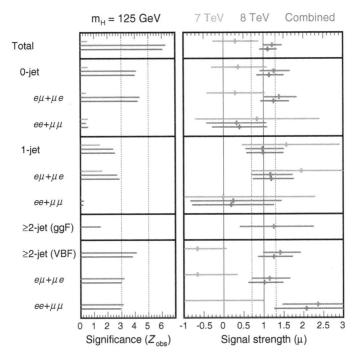

Fig. 8.4 The observed significance in units of Gaussian standard deviations, Z, and the measured signal strength, $\hat{\mu}$, when testing $m_H = 125$ GeV. The contribution of each signal region is shown. The fit results using the $\sqrt{s} = 7$ TeV dataset (*green*), the $\sqrt{s} = 8$ TeV dataset (*red*), and their combination (*blue*) are also separated

particles. When doing this, the VH production mode is assigned the same signal strength as VBF because they are both dominated by bosonic couplings. The resulting limits upon (μ_{ggF}, μ_{VBF}) are shown in Fig. 8.5. It can be seen that both ggF and VBF

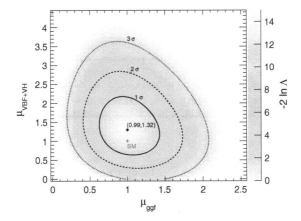

Fig. 8.5 Likelihood contours for μ_{ggF} and $\mu_{\text{VBF+VH}}$, when testing $m_H = 125$ GeV. The markers show the best-fit values observed (*black*) and expected under an $m_H = 125$ GeV hypothesis (*red*)

Table 8.2 A breakdown of the relative uncertainty upon the fitted signal strengths $\hat{\mu}_{\text{ggF}}$ and $\hat{\mu}_{\text{VBF}}$, and in the combined signal strength $\hat{\mu}$

	Relative uncertainty $\Delta\,\mu/\mu$ (%)		
	ggF	VBF	Combined
Statistical	20	33	16
Signal regions	14	29	12
Control regions	12	15	9
Signal contamination	3	7	0
Finite MC size	6	4	4
Systematic	18	21	13
Experimental	10	13	8
Leptons and triggers	5	2	4
Jets and b-tagging	3	10	2
$p_{\text{T}}^{\text{inv}}$ modelling	2	4	2
Fake factor	8	2	5
Other	4	5	3
Theoretical	15	16	11
Signal modelling	11	14	9
Background modelling	9	8	7
Luminosity	3	4	3
Total	26	39	20

"Signal contamination" refers to how the finite number of events in the signal region of the VBF analysis introduces a statistical uncertainty in the expected number of VBF events in the signal region of the ggF analysis, and vice versa. Uncertainties are symmetrised by averaging the high and low variations

are in excellent agreement with the SM, though VBF is observed to be produced more often than expected. The best-fit values are $(\hat{\mu}_{\text{ggF}}, \hat{\mu}_{\text{VBF}}) = (0.99, 1.32)$.

Table 8.2 shows a breakdown of the uncertainties in $\hat{\mu}$. It can be seen that $\hat{\mu}_{\text{VBF}}$ is dominated by statistical uncertainties, whereas $\hat{\mu}_{\text{ggF}}$ is almost equally limited by statistical and systematic uncertainties. It should be noted that theoretical uncertainties

8.3 Results

(particularly in signal modelling) are the largest source of systematic uncertainty, highlighting the importance of the ggF and WW studies in Chaps. 5 and 6 respectively.

8.3.3 Cross Section Measurements

Following the method described in Sect. 5.2.2, fiducial ggF cross sections are extracted for fiducial regions defined for the 0-jet and 1-jet bins of the $e\mu + \mu e$ channel. These are only extracted at $\sqrt{s} = 8$ TeV. This is done using

$$\sigma_{\text{ggF}}^{\text{fid}} = \frac{N_{\text{obs}} - N_{\text{bkg}}}{C_{\text{ggF}} \cdot L} \tag{8.17}$$

$$= \hat{\mu}_{\text{ggF}} \cdot \sigma_{\text{ggF}}^{\text{SM}} \cdot \text{BR}^{\text{SM}} \cdot A_{\text{ggF}} \tag{8.18}$$

where L is the luminosity, C_{ggF} is the ratio of the expected number of ggF events passing the detector-level selection to those passing the fiducial selection, and A_{ggF} is the ratio of the expected number of ggF events passing the fiducial selection to the total expected number of ggF events. Here, $\hat{\mu}$ is determined by a dedicated fit in which the nuisance parameters associated with theoretical uncertainties in the total ggF cross section, the BR and the ggF acceptance are fixed, such that their effect upon $\Delta\mu$ is removed. The VBF and VH processes are set to their SM expectations (i.e. $\mu_{\text{VBF}} = 1$). The measured fiducial cross sections for $m_H = 125$ GeV are displayed in Table 8.3. The measured values are higher than predicted, which is consistent with Fig. 8.4.

The product of the total cross section and the $H \to WW$ branching ratio is extracted for ggF (at $\sqrt{s} = 7$ TeV and 8 TeV) and VBF (at $\sqrt{s} = 8$ TeV) using

$$\sigma_i \cdot \text{BR} = \frac{N_{\text{obs}} - N_{\text{bkg}}}{C_i \cdot A_i \cdot L} \tag{8.19}$$

$$= \hat{\mu}_i \cdot \sigma_i^{\text{SM}} \cdot \text{BR}^{\text{SM}} \tag{8.20}$$

where $i =$ ggF, VBF. The $\hat{\mu}_i$ are determined by a dedicated fit in which the nuisance parameters associated with theoretical uncertainties in the total cross sections and

Table 8.3 Measured fiducial ggF cross sections at $\sqrt{s} = 8$ TeV, assuming $m_H = 125$ GeV.

Fiducial region	σ^{fid} (fb)	
	Measured	Predicted
0-jet $e\mu + \mu e$	28.1±5.5 ± 4.1	19.9± 3.3
1-jet $e\mu + \mu e$	8.4 ± 3.1 ± 1.9	7.3 ± 1.8

Theoretical predictions are shown for comparison. The uncertainties in measured quantities are statistical and systematic, respectively

Table 8.4 Measured ggF and VBF total cross sections multiplied by the $H \rightarrow WW$ branching ratio, assuming $m_H = 125$ GeV

Process	$\sqrt{s}$	$\sigma \cdot$ BR (pb)	
		Measured	Predicted
ggF	7 TeV	1.7±1.8 ± 1.2	3.3± 0.4
ggF	8 TeV	4.7 ± 0.8 ± 0.8	4.1± 0.5
VBF	8 TeV	0.53 ± 0.16 ± 0.10	0.339 ± 0.017

Theoretical predictions are shown for comparison. The uncertainties in measured quantities are statistical and systematic, respectively

branching ratio are fixed, such that their effect is removed. When $\hat{\mu}_{ggF}$ is measured μ_{VBF} is profiled, and vice versa. The measured cross sections for $m_H = 125$ GeV are displayed in Table 8.4 and show good agreement with the theoretical predictions.

References

1. ATLAS Collaboration, Jet energy resolution in proton-proton collisions at $\sqrt{s} = 7$ TeV recorded in 2010 with the ATLAS detector. Eur. Phys. J. C **73**, 2306, (2013). arXiv:1210.6210 [hep-ex]
2. ATLAS Collaboration, Calibration of b-tagging using dileptonic top pair events in a combinatorial likelihood approach with the ATLAS experiment, ATLAS-CONF-2014-004 (2014) http://cds.cern.ch/record/1664335
3. ROOT Collaboration, K. Cranmer, G. Lewis, L. Moneta, A. Shibata, and W. Verkerke, HistFactory: a tool for creating statistical models for use with RooFit and RooStats, CERN-OPEN-2012-016 (2012) http://cds.cern.ch/record/1456844
4. G. Cowan, K. Cranmer, E. Gross, O. Vitells, Asymptotic formulae for likelihood-based tests of new physics. Eur. Phys. J. C **71**, 1554 (2011). arXiv:1727 [physics.data-an]
5. K. Cranmer, Practical Statistics for the LHC, in *7th CERN-Latin American School of High-Energy Physics*, (Arequipa, Peru, 2013)
6. T. Junk, Confidence level computation for combining searches with small statistics. Nucl. Instrum. Methods A **2**(3), 434–435 (1999)
7. A. L. Read, Presentation of search results: the CL_s technique. J. Phys. G: Nucl. Part. Phys. (2002), **28**, 2693
8. ATLAS Collaboration, Search for a high-mass Higgs boson in the $H \rightarrow WW \rightarrow \ell\nu\ell\nu$ decay channel with the ATLAS detector using 21 fb^{-1} of proton-proton collision data, ATLAS-CONF-2013-067 (2013) http://cds.cern.ch/record/1562879
9. ATLAS Collaboration, Observation and measurement of Higgs boson decays to WW^* with the ATLAS detector (2014), in preparation for Phys. Rev. D
10. CMS Collaboration, Measurement of Higgs boson production and properties in the WW decay channel with leptonic final states. (2014). arXiv:1312.1129 [hep-ex]

Chapter 9
Status of Higgs Physics

The analysis presented in this thesis has found significant experimental evidence for the process $gg \to H \to WW \to \ell\nu\ell\nu$. However, it was on 4th July 2012 that the ATLAS and CMS collaborations independently announced the discovery of a new particle consistent with the Higgs boson of the Standard Model (SM), using a combination of decay channels to reject the null hypothesis with more than 5σ significance [1, 2]. Further pp collisions were recorded until the end of December 2012, and then the entire Run I dataset was used to study the new particle in detail.

Section 9.1 summarises the most important measurements of this new particle, which confirm that it has the qualitative properties of the Higgs boson of the SM. The theoretical implications of the discovery are considered in Sect. 9.2. Finally, in Sect. 9.3, the outlook of Higgs boson measurements is assessed in the short and long term future.

9.1 Properties of the Discovered Higgs Boson

For a particle to be considered "discovered" within the particle physics community, the null hypothesis must be rejected with $\geq 5\sigma$ significance, corresponding to $<0.0001\,\%$ chance that the observation is a statistical fluctuation. In the summer of 2012, this level of significance was reached independently by the ATLAS and CMS collaborations, primarily driven by searches for $H \to \gamma\gamma$ and $H \to ZZ$ and supported by the $H \to WW$ search [1, 2]. Both experiments observed a resonance at $m \approx 125\,\text{GeV}$.

This observation was quickly corroborated by $VH \to Vb\bar{b}$ searches performed by the CDF and DØ experiments at the Tevatron, which yielded consistent results with 3.1σ combined significance [3].

Using the entire Run I dataset, the significance of the particle discovery was extended by the ATLAS experiment to 5.2σ in $H \to \gamma\gamma$ [4], 8.1σ in $H \to ZZ$ [5], and 6.1σ in $H \to WW$ (see Chap. 8). Focus inevitably shifted to measuring its properties (mass, couplings, spin, parity) in order to test the hypothesis that it is a Higgs boson. Observations confirm this hypothesis, though more precise measurements are required to ascertain whether or not it is the Higgs boson predicted by the SM.

9.1.1 Mass Measurement

The mass of the Higgs boson, m_H, is unpredicted by the SM, and is very important to measure precisely. Once m_H is known, all the Higgs boson couplings, production cross sections and branching ratios can be calculated (see Sect. 1.3). The value of m_H is also important when assessing the theoretical implications of the discovery (see Sect. 9.2).

The $H \to \gamma\gamma$ and $H \to ZZ$ decay channels offer the best sensitivity to m_H, since they exhibit very clean experimental signatures and allow full reconstruction of the Higgs boson four-momentum. A combination of these two channels results in a measured mass of 125.36 ± 0.37 (stat) ± 0.18 (syst) GeV by the ATLAS collaboration [6] and 125.7 ± 0.3 (stat) ± 0.3 (syst) GeV by the CMS collaboration [7]. These are in excellent agreement, although some tension exists between the two ATLAS channels. The Higgs boson is as heavy as an atom of caesium-133.

A Higgs boson with $m_H \approx 125$ GeV is fortuitous, since many different production modes and decay channels should be experimentally accessible at the LHC (see Figs. 1.4 and 1.5). In particular, couplings to both bosons and fermions can be measured, directly testing the two types of mass generation the Higgs field is responsible for in the SM.

9.1.2 Coupling Measurements and Limits

Coupling strengths of the Higgs boson to other SM particles are probed by searching for different experimental signatures, which correspond to specific combinations of production mode and decay channel. For example, the WH production mode with the $H \to WW$ decay mode is sensitive only to the HWW coupling. On the other hand, the ggF production mode and the $H \to \gamma\gamma$ decay channel feature loops, and interpreting results in terms of couplings is less straightforward. Undiscovered massive particles may contribute to such loops, distorting the effective couplings to gluons and photons.

Measurements of $H \to \gamma\gamma$, $H \to ZZ$, $H \to WW$ and $H \to \tau\tau$ were made with the Run I dataset, with signal regions optimised for ggF and VBF production. A measurement of $VH \to Vb\bar{b}$ was also made. Assuming SM couplings, the measured cross sections can be expressed as a signal strength $\mu = \sigma_{\text{meas}}/\sigma_{\text{SM}}$. The signal strengths measured with the ATLAS experiment are consistent with the SM predictions, as shown in Fig. 9.1. Similar measurements were made with the CMS detector, also yielding signal strengths compatible with the SM [8–12].

9.1 Properties of the Discovered Higgs Boson

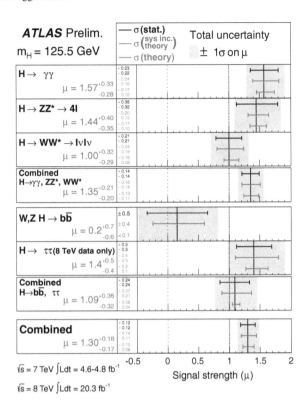

Fig. 9.1 Measured signal strength ($m_H = 125.5\,\text{GeV}$) for each final state [13]. Combinations of bosonic, fermionic and all decay channels are also shown. The $H \to WW$ result is from a previous analysis of the Run I dataset, before the final sensitivity optimisation described in this thesis

It is also informative to split the signal strength according to the production mode. Figure 9.2 shows the likelihood contours for a fit with two signal strength parameters: one for production modes dominated by a fermionic coupling (ggF and ttH), and the other for modes dominated by a bosonic coupling (VBF and VH). Again, this assumes the couplings themselves are those predicted by the SM, but it is useful to see which analyses are sensitive to which types of couplings. It is clear that $H \to WW$ is a very sensitive analysis for coupling measurements.

These measurements were used to test specific coupling scenarios in reference [13], and no significant deviations from the SM were observed. The two most generic scenarios are highlighted here.

The first scenario assumes a SM particle content. The probed couplings are free parameters in the fit, and are expressed as coupling scale factors: κ_Z, κ_W, κ_t, κ_b, κ_τ. The $H\gamma\gamma$ and Hgg effective couplings and the total width are then calculated within the framework of the SM, as functions of the κ_i. Compatibility with the SM is $p = 0.13$ (see Fig. 9.3a), with differences driven by the $H \to b\bar{b}$ result.

The second scenario allows for undiscovered particles, which may contribute through loops or to the total width. Thus, two effective coupling scale factors are introduced (κ_γ, κ_g), and the absence of a total width constraint means that only coupling ratios can be probed. The free parameters of the fit are $\lambda_{\gamma Z}$, λ_{WZ}, λ_{bZ},

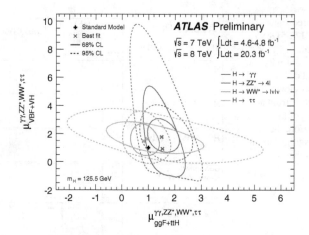

Fig. 9.2 Likelihood contours of measured signal strength ($m_H = 125.5$ GeV) for each final state [13]. Signal strengths are split according to the dominant coupling in the production mode: fermionic (ggF and ttH) or bosonic (VBF and VH). The $H \to WW$ result is from a previous analysis of the Run I dataset

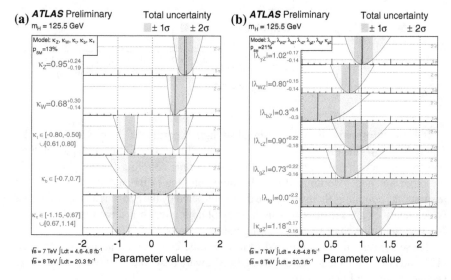

Fig. 9.3 Measured likelihood curves of coupling scale factors κ_i and ratios $\lambda_{ij} = \kappa_i/\kappa_j$, for two generic models described in the text [13]. **a** SM particle content. **b** SM + undiscovered particle content

$\lambda_{\tau Z}$, λ_{gZ}, λ_{tg}, κ_{gZ} where $\lambda_{ij} = \kappa_i/\kappa_j$, $\kappa_{ij} = \kappa_i \cdot \kappa_j/\kappa_H$ and κ_H is the scale factor of the total width. Compatibility with the SM is $p = 0.21$ (see Fig. 9.3b). The large uncertainty on λ_{tg} could be improved by a measurement of the ttH production mode.

9.1 Properties of the Discovered Higgs Boson

In addition to the five measured channels described above, upper bounds on rare decay channels have also been set by ATLAS. The observed limits at $m_H = 125\,\text{GeV}$ are $\sigma < 11\sigma_{\text{SM}}$ for $H \to Z\gamma$ [14] and $\sigma < 7\sigma_{\text{SM}}$ for $H \to \mu\mu$ [15], at the 95 % CL. Also, a search for $ZH \to \ell\ell p_{\text{T}}^{\text{inv}}$ constrained the branching ratio to invisible particles to <75 % at the 95 % CL, assuming $\sigma = \sigma_{\text{SM}}$ and $m_H = 125\,\text{GeV}$ [16]. Such a limit is important because undiscovered massive particles that interact weakly (e.g. dark matter) could enhance the invisible width Γ_{inv}. Slightly tighter limits have been set by the CMS collaboration [17–19].

9.1.3 Spin and Parity Measurement

The SM Higgs boson is a spin-0 and CP-even particle, i.e. $J^P = 0^+$. To establish that the observed particle is the Higgs boson, it is important to experimentally confirm these properties. According to the Landau-Yang theorem, the spin-1 hypothesis is excluded by the observation of $H \to \gamma\gamma$ [20, 21]. Since there may be coincident resonances, this hypothesis is tested by the other channels.

The J^P was probed by considering the decay topologies of events in the $H \to ZZ$, $H \to WW$ and $H \to \gamma\gamma$ measurements with ATLAS. However, some event selection criteria were altered with respect to the coupling measurements, in order to improve sensitivity. For example, the $H \to WW$ selection exploits the spin-0 hypothesis through the $m_{\ell\ell} < 55\,\text{GeV}$ and $\Delta\phi\left(\ell, \ell\right) < 1.8$ criteria (see Sect. 4.3.4).

The $0^-, 1^+, 1^-, 2^+$ hypotheses are excluded at greater than 97.8 % CL, whilst the data are compatible with the 0^+ hypothesis [22]. This provides strong evidence that the discovered particle is spin-0, and is the only observed fundamental scalar particle. It could be a mixture of CP-even and CP-odd states, though a preference for CP-even is observed.

9.2 Theoretical Implications

The discovery of the Higgs boson appears to "complete" the SM in the most minimal way, with no significant deviations from predictions observed. That its HWW and HZZ couplings agree with expectations confirms that the Higgs mechanism underlies electroweak symmetry breaking. That its fermionic couplings are proportional to mass supports that fermion masses are generated by Yukawa interactions.

In Sects. 9.2.1 and 9.2.2, the discovered Higgs boson shall be interpreted assuming that the SM is valid up to the Planck energy scale (at which point gravity becomes strong and dominates phenomenology). Then, Sect. 9.2.3 considers why the observed particle might indicate that new physics should be observed at lower scales.

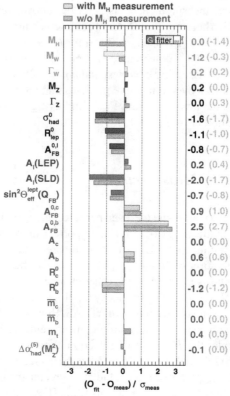

Fig. 9.4 Pull values of the electroweak fit parameters, with (*colour*) and without (*grey*) the m_H measurement included [23]. The pull value is the deviation of the fitted value from the experimental measurement, in units of the experimental uncertainty. With kind permission from Springer Science and Business Media

9.2.1 Global Electroweak Fit

In Sect. 1.4.2, a global fit of electroweak data was used to motivate a low mass Higgs boson. The discovery of the Higgs boson and the measurement of its mass overconstrains the electroweak theory, allowing a test of its validity.

The updated fit exhibits a p-value of 0.176, corresponding to a deviation from the Standard Model of significance 1.35σ [23]. Thus, the experimental data included in the fit are consistent with electroweak theory. The pulls of individual fit parameters are shown in Fig. 9.4; the dominant tension is unrelated to m_H. The measurement of m_H does cause some tension with the measured m_W and m_t, which are sensitive to m_H through loop corrections.

9.2 Theoretical Implications

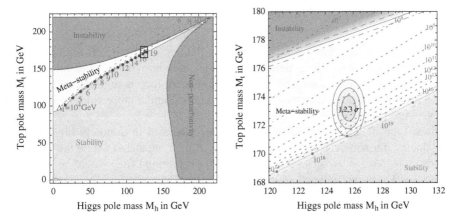

Fig. 9.5 Phase diagram of the Standard Model, expressed in terms of m_H and m_t [25]. The phases correspond to stable, metastable and unstable vacuum states and a non-perturbative Higgs quartic coupling λ, assuming that the scale at which new physics is introduced is $\Lambda_{NP} = \Lambda_P \sim 10^{19}$ GeV. Dotted lines indicate the scale at which the instability phase transition occurs. A zoomed version (*right*) elucidates the experimentally measured situation

9.2.2 Vacuum Stability

In Sect. 1.4.3, theoretical arguments were used to constrain m_H under the assumption that the Standard Model is valid up to the reduced Planck scale $\bar{\Lambda}_P \sim 10^{18}$ GeV. Requiring the Higgs quartic coupling λ remain perturbative implied $m_H < 175$ GeV, while requiring the electroweak vacuum to remain a stable minimum implied $m_H > 129$ GeV [24].

The measurement of $m_H \approx 125$ GeV excludes the stability of the SM vacuum at 98.6 % CL [25]. However, a potential barrier separates the vacuum in which the Universe currently resides and the true SM vacuum. The probability of quantum tunnelling through this barrier is sufficiently small that the lifetime of the Universe far exceeds its age. Thus, measurements suggest that we exist in a metastable vacuum (see Fig. 9.5).

The fact that experimental measurements place the Universe very close to the critical boundary for vacuum stability means that λ and β_λ are both very close to zero at the instability scale. The potential significance of this observation is an active area of theoretical research. One interesting idea also uses the scalar nature of the Higgs boson to lend credence to slow-roll models of cosmic inflation with the Higgs boson acting as the inflaton. However, minimal configurations of such models may fail to predict the power spectrum of anisotropies observed in the cosmic microwave background [26, 27].

9.2.3 The Hierarchy Problem

The Higgs boson acquires a mass through oscillations about the non-zero vacuum expectation value of the Higgs field. However, this tree-level mass is subject to loop corrections containing massive particles; the top quark loop dominates, though the W, Z and Higgs bosons also contribute significantly. Similar corrections to m_W and m_t were used to constrain m_H in Sect. 1.4.2.

Such loop diagrams should in principle be calculated to infinitely high scale, introducing ultraviolet (UV) divergences. They can be calculated within the SM for scales up to $\Lambda_P \sim 10^{19}$ GeV, but a theory of everything (ToE) would be needed above Λ_P. These corrections to m_H are quadratically divergent, and so the measured m_H can be schematically written as

$$m_{H,\text{physical}}^2 = m_{H,\text{bare}}^2 + \int_0^{\Lambda_P} \text{SM loops} + \int_{\Lambda_P}^{\infty} \text{ToE loops} \qquad (9.1)$$

$$= m_{H,\text{bare}}^2 + \mathcal{O}\left(\Lambda_P^2\right) + \int_{\Lambda_P}^{\infty} \text{ToE loops}. \qquad (9.2)$$

The bare mass $m_{H,\text{bare}}$ is not predicted by the SM, but when combined with the ToE loops it must largely cancel the $\mathcal{O}\left(\Lambda_P^2\right)$ term, leaving behind $m_{H,\text{physical}}^2 = (125 \text{ GeV})^2$. This requires the ToE to produce a fine tuning of 1 part in $(\Lambda_P/m_H)^2 \sim 10^{34}$, which is highly unnatural.

One may ask why the hierarchy problem is specific to the Higgs boson, and the same fine tuning is not required for other particles. This is because corrections to the masses of other particles are only logarithmically divergent (rather than quadratically divergent); their masses are protected by gauge or chiral symmetries.

If new physics exists at a scale $\Lambda_{\text{NP}} \ll \Lambda_P$, say $\mathcal{O}(1 \text{ TeV})$, the degree of fine tuning can be reduced dramatically. This is a primary motivation for many new physics models. Supersymmetry introduces a new symmetry between fermions and scalars, which protects m_H [28]. Technicolour regards the Higgs boson as a composite state [29]. Models with extra dimensions assume that gravity is actually strong, but Λ_P appears large because the gravitational flux is diluted in the extra dimensions [30].

Thus, the existence of the Higgs boson motivates new physics at a scale $\mathcal{O}(1 \text{ TeV})$. However, no experimental evidence in support of the candidate models has been found at the LHC, and indeed many configurations have been excluded.

9.3 Outlook

In the short term future, CERN expects to operate Run II of the LHC in 2015 – 2017, and Run III in 2019–2021. Run II is designed to deliver $\sim$100 fb^{-1} at $\sqrt{s} = 13 - 14$ TeV, whilst Run III is designed to deliver $\sim$300 fb^{-1} at $\sqrt{s} = 14$ TeV. In the long term future, CERN plans to further upgrade the LHC instantaneous luminosity (HL-LHC), to start running in 2023 and deliver $\sim$3000 fb^{-1} at $\sqrt{s} = 14$ TeV.[1]

Compared to $\sqrt{s} = 8$ TeV, the Higgs boson production cross sections at $\sqrt{s} = 14$ TeV are $\sim$2.5 times larger for ggF, VBF and VH, and $\sim$4.7 times larger for ttH. Many of the background processes increase by a smaller factor, which should aid the sensitivity of analyses. However, the harsher pile-up environment will degrade detector performance. These considerations, together with the large expected luminosities, should yield more precise signal strength measurements and enable observations of rare decays (e.g. $H \rightarrow Z\gamma$ and $H \rightarrow \mu\mu$) and additional production channels (e.g. ttH). Prospects for the 300 and 3000 fb^{-1} datasets were investigated by ATLAS [31], and the expected precision of signal strengths and coupling scale factors are shown in Fig. 9.6.

The total width is predicted to be $\Gamma_H = 4.07 \pm 0.16$ MeV at $m_H = 125$ GeV [32]. Although this is not experimentally resolvable, it should be possible to indirectly constrain Γ_H at future LHC runs. The Higgs boson propagator $1/((\hat{s}-m_H^2)^2+m_H^2\Gamma_H^2)$ indicates that the on-shell production cross section is sensitive to Γ_H, though it is difficult to disentangle this information from coupling scale factors. However, by comparing the off-shell region to the on-shell region, it is possible to constrain Γ_H with $H \rightarrow ZZ$ and $H \rightarrow WW$ events [33–35]. Also, interference between signal and continuum background can produce a Γ_H-dependent shift in m_H, which could be measured at the HL-LHC with $H \rightarrow \gamma\gamma$ events [36, 37].

At the HL-LHC it should be possible to measure diHiggs production, which would yield a first direct measurement of the Higgs self-coupling λ [38]. This could prove important in understanding the metastability of the vacuum (see Sect. 9.2.2). A measurement of triHiggs production would allow the final Higgs term in the Lagrangian to be measured, though this has an extremely small cross section.

Finally, a solution to the hierarchy problem (see Sect. 9.2.3) shall be sought in future LHC runs. This is a very large topic in itself, though Higgs measurements will play their role in this. Additional (sometimes charged) Higgs bosons often feature in such models, and so searches for these shall continue. Also, it shall be important to further constrain the invisible Higgs width, which can act as a probe of dark matter candidates.

The discovery of the Higgs boson has initiated a new era in high energy physics. It confirms the Higgs mechanism of electroweak symmetry breaking and the mass gen-

---

[1] Another post-LHC scenario is to upgrade the LHC energy to $\sqrt{s} = 33$ TeV (HE-LHC). Highly speculative longer term ideas include building a 100 km tunnel beneath Geneva to house a $\sqrt{s} = 1$ TeV e^+e^- collider (TLEP), and then later a $\sqrt{s} = 100$ TeV pp collider. The international community is also considering proposals for e^+e^- linear colliders, ILC and CLIC, which would have $\sqrt{s} \sim 1$ TeV.

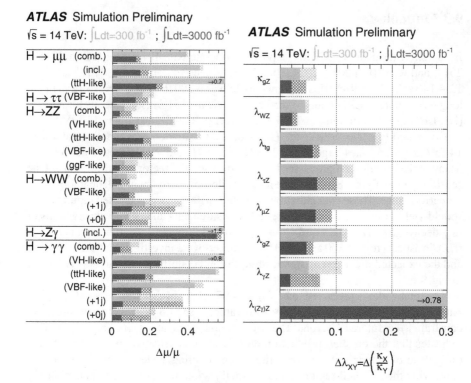

Fig. 9.6 Expected precision of Higgs boson measurements at Run III of the LHC (*green*) and the HL-LHC (*blue*), assuming $m_H = 125\,\text{GeV}$ [31]. The *left* shows signal strength precision for a variety of experimental signatures. The *right* shows the precision of coupling scale factor ratios, following the second generic parametrisation described in Sect. 9.1.2. The hatched areas indicate the decrease in precision due to theoretical uncertainties

eration of fermions via Yukawa interactions, whilst appearing to leave the Standard Model intact and self-consistent. However, it does present interesting opportunities to probe models of new physics.

References

1. ATLAS Collaboration, Observation of a new particle in the search for the Standard Model Higgs boson with the ATLAS detector at the LHC. Phys. Lett. B **716**, 1 (2012). arXiv:1207.7214 [hep-ex]
2. CMS Collaboration, Observation of a new boson at a mass of 125 GeV with the CMS experiment at the LHC. Phys. Lett. B **716**, 30 (2012). arXiv:1207.7235 [hep-ex]
3. CDF, D0 Collaborations, Evidence for a particle produced in association with weak bosons and decaying to a bottom-antibottom quark pair in Higgs boson searches at the tevatron. Phys. Rev. Lett. **109**, 071804 (2012). arXiv:1207.6436 [hep-ex]

4. ATLAS Collaboration, Measurement of Higgs boson production in the diphoton decay channel in pp collisions at center-of-mass energies of 7 and 8 TeV with the ATLAS detector. (2014). arXiv:1408.7084 [hep-ex] (submitted to Phys. Rev. D)
5. ATLAS Collaboration, Measurements of Higgs boson production and couplings in the four-lepton channel in pp collisions at center-of-mass energies of 7 and 8 TeV with the ATLAS detector. (2014). arXiv:1408.5191 [hep-ex] (submitted to Phys. Rev. D)
6. ATLAS Collaboration, Measurement of the Higgs boson mass from the $H \to \gamma\gamma$ and $H \to ZZ^* \to 4\ell$ channels with the ATLAS detector using 25 fb^{-1} of pp collision data. (2014). arXiv:1406.3827 [hep-ex] (accepted by Phys. Rev. D)
7. CMS Collaboration, Combination of standard model Higgs boson searches and measurements of the properties of the new boson with a mass near 125 GeV. CMS-PAS-HIG-13-005 (2013)
8. CMS Collaboration, Observation of the diphoton decay of the Higgs boson and measurement of its properties (2014). arXiv:1407.0558 [hep-ex] (submitted to Eur. Phys. J. C)
9. CMS Collaboration, Measurement of the properties of a Higgs boson in the four-lepton final state, Phys. Rev. D **89**, 092007 (2014). arXiv:1312.5353 [hep-ex]
10. CMS Collaboration, Measurement of Higgs boson production and properties in the WW decay channel with leptonic final states. JHEP **1401**, 096 (2014). arXiv:1312.1129 [hep-ex]
11. CMS Collaboration, Evidence for the 125 GeV Higgs boson decaying to a pair of τ leptons. JHEP **1405**, 104 (2014). arXiv:1401.5041 [hep-ex]
12. CMS Collaboration, Search for the standard model Higgs boson produced in association with a W or a Z boson and decaying to bottom quarks. Phys. Rev. D **89**, 012003 (2014). arXiv:1310.3687 [hep-ex]
13. ATLAS Collaboration, Updated coupling measurements of the Higgs boson with the ATLAS detector using up to 25 fb^{-1} of proton-proton collision data, ATLAS-CONF-2014-009 (2014)
14. ATLAS Collaboration, Search for Higgs boson decays to a photon and a Z boson in pp collisions at $\sqrt{s} = 7$ and 8 TeV with the ATLAS detector. Phys. Lett. B **732**, 8 (2014). arXiv:1402.3051 [hep-ex]
15. ATLAS Collaboration, Search for the Standard Model Higgs boson decay to $\mu^+\mu^-$ with the ATLAS detector (2014). arXiv:1406.7663 [hep-ex] (accepted by Phys. Lett. B)
16. ATLAS Collaboration, Search for invisible decays of a Higgs boson produced in association with a Z boson in ATLAS. Phys. Rev. Lett. **112**, 201802 (2014). arXiv:1402.3244 [hep-ex]
17. CMS Collaboration, Search for a Higgs boson decaying into a Z and a photon in pp collisions at $\sqrt{s} = 7$ and 8 TeV. Phys. Lett. B **726**, 587 (2013). arXiv:1307.5515 [hep-ex]
18. CMS Collaboration, Search for the standard model Higgs boson in the dimuon decay channel in pp collisions at $\sqrt{s} = 7$ and 8 TeV. CMS-PAS-HIG-13-007 (2013)
19. CMS Collaboration, Search for invisible decays of Higgs bosons in the vector boson fusion and associated ZH production modes. Eur. Phys. J. C **74**, 2980 (2014). arXiv:1404.1344 [hep-ex]
20. L.D. Landau, On the angular momentum of a two-photon system. Dokl. Akad. Nauk Ser. Fiz. **60**, 207 (1948)
21. C.-N. Yang, Selection rules for the dematerialization of a particle into two photons. Phys. Rev. **77**, 242 (1950)
22. ATLAS Collaboration, Evidence for the spin-0 nature of the Higgs boson using ATLAS data. Phys. Lett. B **726**, 120 (2013). arXiv:1307.1432 [hep-ex]
23. M. Baak et al., The electroweak fit of the standard model after the discovery of a new boson at the LHC. Eur. Phys. J. C **72**, 1 (2012). arXiv:1209.2716 [hep-ph], updated results taken from http://cern.ch/gfitter (Sep 13)
24. J. Ellis, J.R. Espinosa, G.F. Giudice, A. Hoecker, A. Riotto, The probable fate of the standard model. Phys. Lett. B **679**, 369 (2009). arXiv:0906.0954 [hep-ph]
25. D. Buttazzo et al., Investigating the near-criticality of the Higgs boson. JHEP **1312**, 089 (2013). arXiv:1307.3536
26. G. Isidori, V.S. Rychkov, A. Strumia, N. Tetradis, Gravitational corrections to standard model vacuum decay. Phys. Rev. D **77**, 025034 (2008). arXiv:0712.0242 [hep-ph]
27. A. De Simone, M.P. Hertzberg, F. Wilczek, Running inflation in the standard model. Phys. Lett. B **678**, 1 (2009). arXiv:0812.4946 [hep-ph]

28. S.P. Martin, A Supersymmetry Primer (1997). arXiv:hep-ph/9709356
29. M.E. Peskin, Beyond the Standard Model (1997). arXiv:hep-ph/9705479
30. A. Pomarol, Beyond the Standard Model (2012). arXiv:1202.1391 [hep-ph]
31. ATLAS Collaboration, Projections for measurements of Higgs boson cross sections, branching ratios and coupling parameters with the ATLAS detector at a HL-LHC, ATL-PHYS-PUB-2013-014 (2013)
32. LHC Higgs Cross Section Working Group, Handbook of LHC Higgs Cross Sections: 3. Higgs Properties, CERN-2013-004 (2013). arXiv:1307.1347 [hep-ph]
33. F. Caola, K. Melnikov, Constraining the Higgs boson width with ZZ production at the LHC. Phys. Rev. D **88**, 054024 (2013). arXiv:1307.4935 [hep-ph]
34. J.M. Campbell, R.K. Ellis, C. Williams, Bounding the Higgs width at the LHC using full analytic results for $gg \to e^-e^+\mu^-\mu^+$. JHEP **1404**, 060 (2014). arXiv:1311.3589 [hep-ph]
35. J.M. Campbell, R.K. Ellis, C. Williams, Bounding the Higgs width at the LHC: complementary results from $H \to WW$. Phys. Rev. D **89**, 053011 (2014). arXiv:1312.1628 [hep-ph]
36. L.J. Dixon, Y. Li, Bounding the Higgs boson width through interferometry. Phys. Rev. Lett. **111**, 111802 (2013). arXiv:1305.3854 [hep-ph]
37. S.P. Martin, Interference of Higgs diphoton signal and background in production with a jet at the LHC. Phys. Rev. D **88**, 013004 (2013). arXiv:1303.3342 [hep-ph]
38. M.J. Dolan, C. Englert, M. Spannowsky, Higgs self-coupling measurements at the LHC. JHEP **1210**, 112 (2012). arXiv:1206.5001 [hep-ph]

Chapter 10
Conclusions

This thesis has described the experimental search for the $gg \to H \to WW \to \ell\nu\ell\nu$ process of Higgs boson production and decay. It uses the LHC Run I dataset of pp collisions recorded by the ATLAS detector, which corresponds to an integrated luminosity of 4.5 fb^{-1} at $\sqrt{s} = 7$ TeV and 20.3 fb^{-1} at $\sqrt{s} = 8$ TeV. An excess of events is observed with a significance of 4.8 standard deviations (4.8σ), which is consistent with Higgs boson production. The significance is extended to 6.1σ when the vector boson fusion production process is included. According to the convention adopted by the particle physics community, this constitutes a first observation, or discovery, of this process. The observed resonance is found to be consistent with the Higgs boson of the Standard Model with $m_H = 125$ GeV, as are results from other LHC search channels described in Sect. 9.1.

The best-fit signal strength at $m_H = 125$ GeV is found to be $\hat{\mu} = 1.11 \pm 0.22$, in excellent agreement with the Standard Model expectation. With a precision of 20 %, this $H \to WW$ analysis is the most sensitive μ measurement of the LHC Run I Higgs boson analyses [1–5]. It will take some years to improve upon this precision at Run II of the LHC. With the expected Run II dataset of 100 fb^{-1} at $\sqrt{s} = 13$–14 TeV, the statistical uncertainty should be reduced – dramatically so for μ_{VBF}. The larger dataset should also afford more sophisticated background estimation techniques to be employed, e.g. using same-sign events to model the normalisation *and* shape of the non-WW diboson background, in order to reduce the associated theoretical uncertainties. Finally, advances in theoretical calculations and MC event generators shall improve the estimation of processes with large theoretical uncertainties (e.g. ggF and WW).

References

1. ATLAS Collaboration, Measurement of Higgs boson production in the diphoton decay channel in pp collisions at center-of-mass energies of 7 and 8 TeV with the ATLAS detector (2014). arXiv:1408.7084 [hep-ex] (submitted to Phys. Rev. D)
2. ATLAS Collaboration, Measurements of Higgs boson production and couplings in the four-lepton channel in pp collisions at center-of-mass energies of 7 and 8 TeV with the ATLAS detector (2014). arXiv:1408.5191 [hep-ex] (submitted to Phys. Rev. D)
3. CMS Collaboration, Observation of the diphoton decay of the Higgs boson and measurement of its properties (2014). arXiv:1407.0558 [hep-ex] (submitted to Eur. Phys. J. C)
4. CMS Collaboration, Measurement of the properties of a Higgs boson in the four-lepton final state. Phys. Rev. D **89**, 092007 (2014). arXiv:1312.5353 [hep-ex]
5. CMS Collaboration, Measurement of Higgs boson production and properties in the WW decay channel with leptonic final states. JHEP **1401**, 096 (2014). arXiv:1312.1129 [hep-ex]

About the Author

David Hall obtained a first class MPhys degree at the University of Warwick in 2010. For his masters research, he developed a prototype for a novel intensity modulated radiotherapy treatment modality at the University Hospital in Coventry, under the supervision of Prof. Adrian Wilson.

David received his DPhil degree in 2014 at the University of Oxford whilst working on the ATLAS experiment at CERN. This involved measuring the *WW* cross section and searching for evidence of the Higgs boson, under the supervision of Dr Chris Hays. He then worked as the ATLAS Monte Carlo software coordinator in the preparation for Run-II of the LHC.

In January 2015, David moved into proton therapy research in order to combine his previous research experiences, and spent a short time at the Particle Therapy

Cancer Research Institute, University of Oxford. He is now a postdoctoral research fellow at Massachusetts General Hospital and Harvard Medical School. He develops Monte Carlo simulation programs and treatment planning tools for proton therapy, in Prof. Harald Paganetti's group.

Selected Publications

1. ATLAS Collaboration, Observation and measurement of Higgs boson decays to WW^* with the ATLAS detector (2014). arXiv:1412.2641 [hep-ex] (submitted to Phys. Rev. D)
2. ATLAS Collaboration, Measurements of Higgs boson production and couplings in diboson final states with the ATLAS detector at the LHC. Phys. Lett. B **726**, 88 (2013). arXiv:1307.1427 [hep-ex]
3. LHC Higgs Cross Section Working Group, Handbook of LHC Higgs Cross Sections: 3. Higgs Properties. CERN-2013-004 (2013). arXiv:1307.1347 [hep-ph]
4. ATLAS Collaboration, A particle consistent with the Higgs boson observed with the ATLAS detector at the Large Hadron Collider. Science **338**, 1576 (2012)
5. ATLAS Collaboration, Observation of a new particle in the search for the standard model Higgs boson with the ATLAS detector at the LHC. Phys. Lett. B **716**, 1 (2012). arXiv:1207.7214 [hep-ex]
6. ATLAS Collaboration, Measurement of W^+W^- production in pp collisions at $\sqrt{s} = 7$ TeV with the ATLAS detector and limits on anomalous WWZ and $WW\gamma$ couplings. Phys. Rev. D **87**, 112001 (2013). arXiv:1210.2979 [hep-ex]
7. ATLAS Collaboration, Measurement of the WW cross section in $\sqrt{s} = 7$ TeV pp collisions with the ATLAS detector and limits on anomalous gauge couplings. Phys. Lett. B **712**, 289 (2012). arXiv:1203.6232 [hep-ex]
8. D.C. Hall, P. Hamilton, B.T. Huffman, P.K. Teng, A.R. Weidberg, The radiation tolerance of MTP and LC optical fibre connectors to 500 kGy(Si) of gamma radiation. JINST **7**, P04014 (2012)
9. D. Hall, B.T. Huffman, A. Weidberg, The radiation induced attenuation of optical fibres below $-20\,°C$ exposed to lifetime HL-LHC doses at a dose rate of 700 Gy(Si)/hr. JINST **7**, C01047 (2012)
10. F. Vasey et al., The Versatile Link common project: feasibility report. JINST **7**, C01075 (2012)

CPSIA information can be obtained
at www.ICGtesting.com
Printed in the USA
LVHW02*1214300918
591916LV00004B/590/P